U.S. Department
of Transportation

**National Highway
Traffic Safety
Administration**

I0470858

DOT HS 810 748

NHTSA Technical Report

January 2007

An Evaluation of Side Impact Protection

FMVSS 214 TTI(d) Improvements and Side Air Bags

This document is available to the public at the Docket Management System of the U.S. Department of Transportation.

The United States Government does not endorse products or manufacturers. Trade or manufacturers' names appear only because they are considered essential to the object of this report.

1. Report No. DOT HS 810 748	2. Government Accession No.	3. Recipient's Catalog No.
4. Title and Subtitle AN EVALUATION OF SIDE IMPACT PROTECTION FMVSS 214 TTI(d) Improvements and Side Air Bags		5. Report Date January 2007
		6. Performing Organization Code
7. Author(s) Charles J. Kahane, Ph.D.		8. Performing Organization Report No.
9. Performing Organization Name and Address Evaluation Division; National Center for Statistics and Analysis National Highway Traffic Safety Administration Washington, DC 20590		10. Work Unit No. (TRAIS)
		11. Contract or Grant No.
12. Sponsoring Agency Name and Address Department of Transportation National Highway Traffic Safety Administration Washington, DC 20590		13. Type of Report and Period Covered NHTSA Technical Report
		14. Sponsoring Agency Code
15. Supplementary Notes		

16. Abstract

Federal Motor Vehicle Safety Standard 214, "Side Impact Protection" was amended to assure occupant protection in a 33.5 mph crash test and phased-in to new passenger cars during model years 1994-1997. A Thoracic Trauma Index, TTI(d) is measured on Side Impact Dummies seated adjacent to the impact point. Manufacturers upgraded side structures and affixed padding in cars to improve TTI(d). Later, they installed two types of side air bags – torso bags and head air bags – for additional occupant protection in cars and LTVs. Statistical analyses of 1993-2005 crash data from the Fatality Analysis Reporting System (FARS) and the General Estimates System (GES) estimate fatality reductions for these technologies.

- Average TTI(d) improved in 2-door cars from 114 in 1981-1985 to 44 in 214-certified cars with side air bags, and in 4-door cars from 85 to 48.

- TTI(d) improvements without side air bags reduced fatality risk for nearside occupants in multivehicle crashes by an estimated 33 percent in 2-door cars and 17 percent in 4-door cars.

- Torso plus head air bags reduce fatality risk for nearside occupants by an estimated 24 percent; torso bags alone, by 12 percent.

- TTI(d) improvements, torso bags and head-curtain air bags could have saved an estimated 2,934 lives in calendar year 2003 if every car and LTV on the road had been equipped with them.

17. Key Words NHTSA; FARS; NASS; GES; FMVSS; statistical analysis; evaluation; benefits; effectiveness; fatality reduction; crashworthiness; angle collision	18. Distribution Statement Document is available to the public at the Docket Management System of the U.S. Department of Transportation, http://dms.dot.gov, Docket Number 27588.		
19. Security Classif. (Of this report) Unclassified	20. Security Classif. (Of this page) Unclassified	21. No. of Pages 178	22. Price

Form DOT F 1700.7 (8-72)　　　　Reproduction of completed page authorized

TABLE OF CONTENTS

LIST OF ABBREVIATIONS

ABS   Antilock brake system

AIS   Abbreviated Injury Scale

ANPRM   Advance Notice of Proposed Rulemaking

BMW   Bayerische Motoren Werke

CATMOD   Categorical models procedure in SAS

df   Degrees of freedom

DOT   United States Department of Transportation

EU   European Union

FARS   Fatality Analysis Reporting System (a census of fatal crashes in the United States since 1975)

FMH   Free-motion headform

FMVSS   Federal Motor Vehicle Safety Standard

GES   General Estimates System of NASS

GM   General Motors

GVWR   Gross Vehicle Weight Rating (specified by the manufacturer, equals the vehicle's curb weight plus maximum recommended loading)

HIC   Head Injury Criterion

IIHS   Insurance Institute for Highway Safety

IR   Information Request (from NHTSA)

kph   Kilometers per hour

LTV   Light trucks and vans (includes pickup trucks, SUVs, minivans and full-sized vans)

MDB   Moving deformable barrier

mph   Miles per hour

msec	Milliseconds
MY	Model year
NASS	National Automotive Sampling System (a probability sample of police-reported crashes in the United States since 1979, investigated in detail)
NCAP	New Car Assessment Program (consumer information supplied by NHTSA on the safety of new cars and LTVs, based on test results, since 1979)
NHTSA	National Highway Traffic Safety Administration
NICB	National Insurance Crime Bureau
NPRM	Notice of Proposed Rulemaking
NVPP	R.L. Polk's National Vehicle Population Profile
PSU	Primary sampling unit
RF	Right-front
SAS	Statistical analysis software produced by SAS Institute, Inc.
SID	Side impact dummy
SUV	Sport utility vehicle
TTI	Thoracic Trauma Index
TTI(d)	Thoracic Trauma Index for the dummy in a side-impact test
VIN	Vehicle Identification Number
VMT	Vehicle miles of travel

ACKNOWLEDGEMENTS

I owe special thanks to the three researchers who peer-reviewed a draft of this report:

1) Mr. Dainius J. Dalmotas, Chief, Crashworthiness Research, Motor Vehicle Standards and Research, Transport Canada (Retired), Gatineau, Quebec, Canada

2) Mr. John L. Jacobus, Mechanical Engineer, National Highway Traffic Safety Administration (Retired), Silver Spring, MD

3) Mr. Anders Lie, Vehicle Safety Specialist, Swedish National Road Administration, Borlange, Sweden

This study estimates the fatality-reducing effectiveness of side air bags, based on statistical analyses of crash data. The National Highway Traffic Safety Administration (NHTSA) plans to use the statistical results when it estimates the benefits of a future final rule: the addition of a pole test to Federal Motor Vehicle Safety Standard (FMVSS) No. 214, Side Impact Protection. Because of the potential cost impacts of the proposed regulation, the report contains "highly influential scientific information" as defined by the Office of Management and Budget's (OMB) "Final Information Quality Bulletin for Peer Review" (available at www.whitehouse.gov/omb/inforeg/peer2004/peer_bulletin.pdf). Therefore, the report had to be peer-reviewed in accordance with the requirements of both Sections II and III of OMB's Bulletin.

The peer-review process differed from the type used by journals. The effort by Messrs. Dalmotas, Jacobus and Lie was essentially consultation to identify shortcomings in the draft and help NHTSA strengthen the report. We in NHTSA specifically requested and arranged for these three reviewers. The review process is on record – their comments on the draft may be viewed in the NHTSA docket for this report. The publication of this report does not necessarily imply that they "endorsed" it or agreed with its findings. You may read their comments in the docket to see what they agreed or disagreed with in the draft. We have tried to address all of the comments in our revised report (but we did not send it back to them for a second round of review). The text and footnotes of the report single out some of the reviewers' comments that instigated additions or revisions to the analyses.

EXECUTIVE SUMMARY

Side air bags with head protection, such as torso bags with head curtains reduce fatality risk in side impacts by an estimated 24 percent for the nearside occupant, the person seated adjacent to the struck side of the vehicle. That benefit adds to the effect of improved side structures and padding built into passenger cars during the 1980s and 90s that had already reduced fatality risk for nearside occupants by 33 percent in 2-door cars and 17 percent in 4-door cars.

In 2003, over 9,000 fatalities, approximately 29 percent of all occupant fatalities in cars and LTVs (light trucks and vans – i.e., pickup trucks, sport utility vehicles, minivans and full-size vans) began with a side impact. The side of a vehicle, especially the door area adjacent to the occupant is intrinsically a vulnerable spot: there is limited space and structure between the occupant and the outside. Side impacts can also be difficult to avoid. Even the most prudent driving on our part cannot eliminate the risk that another vehicle will fail to yield, run a red light or turn without warning across our path.

Since the 1970's, the National Highway Traffic Safety Administration (NHTSA), the manufacturers and others in the safety community have worked hard to reduce fatality risk in side impacts, especially for the most vulnerable occupant, the "nearside" occupant: the driver in a left-side impact and the right-front passenger in a right-side impact. The effort resulted in the four tangible improvements in side impact protection that are evaluated in this report:

1. Upgrading the side **structure** of passenger cars to slow down and reduce the extent of door intrusion into the passenger compartment after a side impact. Improvements include redesigning or strengthening the beams that horizontally reinforce the doors; the pillars, sills, and roof rails that surround the doors; and the cross-members or seat structures that resist lateral crush.

2. Installation of thick, energy absorbing **padding** within the door structure to reduce the probability of occupant injury after the door interior contacts the occupant.

And two types of **side air bags**:

3. **Torso air bags** that deploy from the seat or the door to provide an energy-absorbing cushion between the occupant's torso and the vehicle's side structure. Torso air bags cover a much larger impact area and absorb more energy than padding.

4. **Head-protection air bags** that complement the torso bags by cushioning head impacts with the side structure and possibly barring occupant ejection through side windows. Head protection may consist of:

    a. "Torso/head combination bags" that deploy from the seat to protect the torso but also extend upward far enough to protect the head impact zones around the side window, or

    b. "Head curtains" or "inflatable tubular structures" that drop down from the roof rail into the side-window area, separately from the torso bags.

During the 1980's, NHTSA and the safety community developed a procedure for assessing injury risk in side impacts, including:

- A crash test configuration simulating a severe intersection collision in which a fast-moving vehicle strikes a slow-moving vehicle in the door, at a right angle.

- A Moving Deformable Barrier (MDB) simulating a generic striking vehicle.

- A Thoracic Trauma Index (TTI) that predicts the severity of thoracic injuries when occupants' torsos contact the interior side surface of the struck vehicle.

- A Side Impact Dummy (SID) on which TTI can be reliably measured in side impact tests. The injury score measured on the dummy is called TTI(d).

In 1990 NHTSA amended Federal Motor Vehicle Safety Standard (FMVSS) 214, *Side Impact Protection* for passenger cars, adding a 33.5 mph impact by an MDB into the side of the car and limiting TTI(d) for a SID in the nearside position up to a maximum of 90 in 2-door cars and 85 in 4-door cars. The requirement was phased-in to passenger cars during model years 1994 to 1997 and subsequently extended to LTVs, effective in model year 1999, limiting TTI(d) to 85.

The manufacturers redesigned structures and/or affixed padding to substantially reduce average TTI(d) during and, to some extent, even before the 1994-1997 phase-in of FMVSS 214. But their actions varied from model to model. Many 2-door cars, with their long, vulnerable door areas, received extensive structural reinforcement or other redesign, whereas some of the heavier 4-door cars and most LTVs needed little or no change to meet FMVSS 214. In many cars, manufacturers improved TTI(d) well beyond the NHTSA requirements.

Manufacturers have continued to improve side impact protection by installing side air bags and/or upgrading side structures as they redesigned their cars. Torso bags first appeared on production vehicles in 1996 and head-protection air bags in 1998. By model year 2003, nearly 30 percent of new cars were equipped with torso bags and nearly 20 percent with head-protection air bags. NHTSA does not require side air bags, but encourages all improvements to side impact protection, including side air bags, by informing consumers about the performance of new vehicles. The agency's New Car Assessment Program (NCAP) includes a rating system of one star (worst) to five stars (best) on a side impact test. *Buying a Safer Car* brochures specify what make-models are equipped with torso and/or head air bags. The information is available to consumers on the agency's web site, www.safercar.gov.

TTI(d) performance at the 33.5 mph test speed of FMVSS 214 demonstrates how much cars have improved over the years. In 2-door cars, TTI(d) for front-seat occupants has improved, on the average, from 114 in baseline 1981-1985 models to 44 in models equipped with side air bags and meeting FMVSS 214: amazing progress on a difficult safety problem.

This report investigates if the improvements in side impact protection have saved lives in actual crashes, based on statistical analyses of crash data. The Government Performance and Results Act of 1993 and Executive Order 12866 require agencies to evaluate the benefits of their existing regulations. The statistical analyses use calendar year 1993-2005 crash data from the Fatality

Analysis Reporting System (FARS) and the General Estimates System (GES) of the National Automotive Sampling System (NASS). The analyses are divided into two main sections:

- Effect of TTI(d) improvements by structure and padding (without side air bags) on the fatality risk of front-seat occupants (drivers and right-front passengers) in passenger cars. Many of the improvements date to the mid-1990s. By now, the cars have been on the road for nearly a decade. While there is a fair amount of uncertainty, the results are essentially final in the sense that most of the eventual data are already in hand.

    o A parallel analysis for compact pickup trucks did not show a statistically significant effect.

- Effect of side air bags – torso bags and/or head-protection air bags – for front-seat occupants of cars and LTVs. Side air bags, especially head air bags began to appear in large numbers only after 2000. Analyses already show statistically significant results, but more data are on the way. The findings of this report will be updated periodically during the next five years.

    o Side air bags are principally designed to protect nearside occupants but might conceivably also benefit farside occupants: the driver in a right-side impact and the right-front passenger in a left-side impact. Statistical analyses separately focus on nearside and farside occupants.

The main findings of this report are that structural improvements and padding for cars, and side air bags for cars and LTVs have significantly reduced occupants' fatality risk. The two types of side air bags – torso bags and head-protection air bags – make substantial and complementary contributions to fatality reduction for nearside occupants. Head curtains (or inflatable tubular structures) also appear to have a significant benefit for farside occupants of passenger cars. The public will obtain the most protection if they have all of these improvements: structures and padding that meet or exceed the requirements of FMVSS 214, torso bags and head curtains. The combined effects are impressive, amounting to a 42 percent cumulative fatality reduction in 2-door cars, and a 30 percent reduction in 4-door cars.

The findings and conclusions of the statistical analyses are the following:

SIDE IMPACT PERFORMANCE OVER THE YEARS

The risk of chest injury in a side impact is measured on a specially designed side impact dummy during a crash test in the FMVSS 214 configuration, a 33.5 mph impact by a moving deformable barrier into the side of the test vehicle. Accelerations measured on the upper and lower ribs and lower spine are combined into a Thoracic Trauma Index for the dummy - TTI(d). TTI(d) gauges occupants' injury risk in nearside impacts: the lower the TTI(d), the lower the risk of injury. Reductions in the average TTI(d) of the many vehicles NHTSA has tested over the years demonstrate improved safety in side impacts.

- TTI(d) for front-seat occupants in the FMVSS 214 test configuration, by model year, averaged:

	2-Door Cars	4-Door Cars
FMVSS 214 requirement	*90*	*85*
Actual performance:		
1981-1985 baseline TTI(d)	114	85
1993-1996, but not yet 214 certified	95	71
1994-2003, 214-certified – no side air bags	69	63
1996-2003, 214-certified – with side air bags	44	48

- In 2-door cars, TTI(d) improved by 45 units since 1981-1985 without side air bags and an additional 25 units with side air bags, for a total of 70. Average performance was originally much worse than the FMVSS 214 requirement and is now much better.

- In 4-door cars, TTI(d) improved by 22 units since 1981-1985 without side air bags and an additional 15 units with side air bags, for a total of 37. Average performance was once about the same as the FMVSS 214 requirement and is now much better.

- TTI(d) performance used to be much worse in 2-door cars than in 4-door cars; it is now nearly the same.

EFFECT OF TTI(d) IMPROVEMENT WITHOUT SIDE AIR BAGS IN PASSENGER CARS

- During the model year 1994-1997 phase-in of FMVSS 214, approximately:

  - 56 percent of cars received substantial structural modifications, usually accompanied with padding.

  - 21 percent received padding with minor structural modifications.

  - 6 percent received padding only.

  - 17 percent remained essentially unchanged from previous model years.

- This report identifies 15 make-models that substantially improved TTI(d), by a known amount, without side air bags: from an average of 85 to 62, a 23-unit improvement. Fatality risk of nearside front-seat occupants in multivehicle crashes decreased by a statistically significant 18 percent in these models (90 percent confidence bounds, 7 to 28 percent).

- For passenger cars with TTI(d) in the below-90 range, each unit improvement of TTI(d) without side air bags is associated with an estimated 0.863 percent fatality reduction for nearside occupants in multivehicle crashes (confidence bounds, 0.33 to 1.46 percent).

  - The fatality reductions for nearside occupants in single-vehicle crashes and for farside occupants were not statistically significant.

- For pre-FMVSS 214, 2-door cars with TTI(d) in the 90+ range, each unit improvement of TTI(d) was associated with an estimated 0.927 percent fatality reduction for all occupants in side impacts (confidence bounds, 0.52 to 1.33 percent).

- In 2-door cars, the cumulative effect of reducing TTI(d) from 114 (1981-1985 baseline) to 69 (post-FMVSS 214 without side air bags) is a 33 percent fatality reduction for nearside occupants in multivehicle crashes (confidence bounds, 18 to 47 percent).

- In 4-door cars, the cumulative effect of reducing TTI(d) from 85 (1981-1985 baseline) to 63 (post-FMVSS 214 without side air bags) is a 17 percent fatality reduction for nearside occupants in multivehicle crashes (confidence bounds, 7 to 27 percent).

- TTI(d) improvement by structures and padding in passenger cars saved an estimated 803 lives in calendar year 2003.

- If every passenger car on the road in 2003 had been equipped with these improvements, they would have saved an estimated 1,143 lives.

EFFECT OF SIDE AIR BAGS IN CARS AND LTVs

Nearside occupants

- Torso bags plus head protection in passenger cars reduces the fatality risk of nearside front-seat occupants in single- and multivehicle crashes by a statistically significant 24 percent (90 percent confidence bounds, 4 to 42 percent).[1]

  o The data also show a statistically significant fatality reduction in LTVs and suggest that the effectiveness may be the same as in cars.

  o The available data do not show a difference in fatality reduction between the two types of head air bags: head curtains (or inflatable tubular structures) and torso/head combination bags.

- Torso bags alone reduce the fatality risk of nearside occupants in passenger cars by an estimated 12 percent (confidence bounds, -3 to +23 percent).

  o Current data also suggest similar reductions for LTV occupants.

- Through 2005, there were few vehicles equipped with head curtains only (no torso bags): not enough for a separate statistical analysis. However, the preceding results suggest that torso bags and head air bags are both effective in nearside impacts and make approximately equal contributions to fatality reduction.

Farside occupants

- Specific mechanisms whereby side air bags mitigate injuries in farside impacts have not yet been widely demonstrated or quantified by testing.

- Nevertheless, statistical analyses of FARS and GES data show significant reductions of fatality risk for head curtains plus torso bags in farside impacts to passenger cars.

- Furthermore, analyses of life-threatening injuries to farside occupants in passenger cars without side air bags suggest that head curtains or inflatable tubular structures could have benefited unrestrained occupants – or even belted drivers if no passenger had been sitting between them and the right side of the car – because:

  o Head curtains would have deployed and covered areas responsible for a large proportion of the life-threatening injuries, and

---

[1] A small portion of this effectiveness may actually be due to energy-absorbing materials (other than air bags) installed to meet the FMVSS 201 upgrade of head-impact protection. NHTSA will evaluate FMVSS 201 in the future; this report only addresses its interaction with side air bags. In many make-models, the introduction of head air bags coincided with FMVSS 201 certification; nevertheless, the energy-absorbing materials remained largely unchanged in the year that head air bags were introduced, and for that reason could not have accounted for a large portion of the fatality reduction for those make-models in that year.

- o In most of those impacts, the head curtains would still have been at least partially inflated at the time the farside occupant contacted them.

- A 24 percent fatality reduction is estimated (same as for nearside occupants) for head curtains plus torso bags in farside impacts to passenger cars – for unrestrained occupants and for belted drivers riding alone in the front seat.

- With the limited crash data available to date, no consistently significant fatality reduction was found and, for now, none is claimed in farside impacts for:

  - o LTVs (with any type of side air bags),

  - o Torso bags alone or torso/head combination bags in cars, or

  - o Belted occupants, when somebody sits between them and the far side.

Occupant ejection

- Head curtains reduced the risk of fatal occupant ejection in side impacts by a statistically significant 30 percent.

  - o Through model year 2003, head air bags in passenger cars were only designed to deploy in side impacts. Head curtains with rollover sensors began to appear in selected LTVs during mid-model year 2002. Crash data were not sufficient to evaluate to what extent this promising technology reduces ejections in rollover crashes.

Overall

- Side air bags could have saved an estimated 1,791 lives in calendar year 2003 if every passenger car and LTV on the road had been equipped with head curtains (or inflatable tubular structures) plus torso bags and if every LTV on the road had been equipped with torso bags plus head protection. However, the number of lives saved if all vehicles on the road were to have side air bags in a future year would be smaller than 1,791, since:

  - o The long-term shift of the on-road fleet from cars to LTVs will reduce the number of potentially fatal side impacts because LTVs are less vulnerable, when struck in the side, than cars.

  - o The increasing proportion of vehicles equipped with Electronic Stability Control will further reduce the number of potentially fatal side impact and rollover crashes by preventing these crashes altogether.

The estimation of future lives saved is beyond the scope of this report, but will be addressed in NHTSA's forthcoming Final Regulatory Impact Analysis to add a pole test to FMVSS 214.

COMBINED EFFECT OF IMPROVED STRUCTURE, PADDING, AND SIDE AIR BAGS

- Side impact protection could have saved an estimated 2,934 lives in calendar year 2003 if every car on the road had been equipped with head curtains, torso bags and FMVSS 214 side structures/padding, and if every LTV on the road had been equipped with torso bags plus head protection.

- Relative to 1981-1985 baseline cars, the combination of head curtains, torso bags and FMVSS 214 side structures/padding reduces fatality risk of drivers and right-front passengers in all side impacts by:

    o 42 percent in 2-door cars.

    o 30 percent in 4-door cars.

- In LTVs, torso bags plus head protection reduce fatality risk of drivers and right-front passengers in all side impacts by 15 percent.

CHAPTER 1

OCCUPANT PROTECTION IN SIDE IMPACTS

Federal Motor Vehicle Safety Standard (FMVSS) 214, amended in 1990 to assure occupant protection in a dynamic test that simulates a side impact collision, is one of the most important safety regulations issued by NHTSA. The requirement was phased-in to passenger cars during model years 1994 to 1997. Crash data are now available to evaluate whether this regulation and the vehicle modifications that improve performance in the side impact test, including upgraded structure, padding and side air bags are effective in reducing fatality risk in actual side impact crashes of production passenger cars.

1.1   The side impact problem in passenger cars

Number of fatalities: Figure 1-1 shows that side impacts accounted for close to 9,000 occupant fatalities per year in passenger cars and LTVs (light trucks and vans, including pickup trucks, SUVs, minivans and full-size vans under 10,000 pounds GVWR), year after year, from 1975 through 2004:[2]

Figure 1-1: Car and LTV Occupant Fatalities in All Side Impacts, 1975-2004

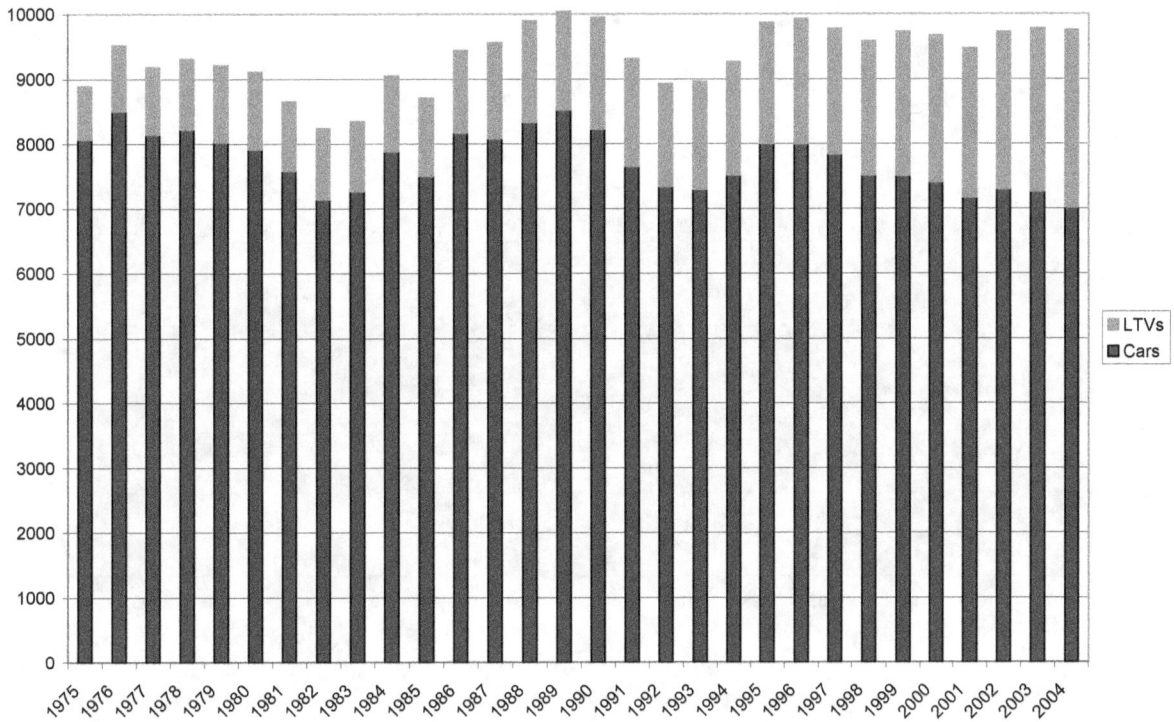

---

[2] Dainius Dalmotas recommended trend analyses of side impact fatalities in his review of this report. Figures 1-1 – 1-9 address issues he raised, but the data in these figures were generated, at NHTSA, especially for this report.

The number of fatalities stayed about the same while vehicle miles of travel (VMT) more than doubled. The proportion of these fatalities in LTVs increased in parallel with the increasing ratio of LTVs to cars in the on-road fleet. Nevertheless, LTVs are under-represented because they are less vulnerable in side impacts than cars. For example, in 2004, only 28 percent of the side impact fatalities were occupants of LTVs, even though LTVs accounted for 40 percent of the on-road fleet.[3]

Figure 1-1 documents that side impacts accounted for 7,000-8,500 occupant fatalities per year in passenger cars throughout 1975-2004; however, they gradually declined from 8,000 to 7,000 in 1996-2004. The decline could reflect the gradual aging of the on-road fleet (older cars are driven fewer miles per year) and also, conceivably, the benefits of safety measures, including the measures evaluated in this report.

Proportion of fatalities: Figure 1-2 shows that side impacts account for a gradually increasing share of the occupant fatalities in passenger cars, rising from 30 percent of the fatalities in 1975 to 37 percent in 2004.

Figure 1-2: Percent of Car Occupant Fatalities that Are in Side Impacts, 1975-2004

---

[3] Overall VMT increased from 1,328 billion miles in 1975 to 2,963 billion miles in 2004; 2740 of 9755 side impact fatalities in 2004 were in LTVs, from Figure 1-1; 89,938,581 of 223,213,958 passenger vehicles registered in 2004 were LTVs, according to *Traffic Safety Facts 2004*, NHTSA Report No. DOT HS 809 919, Washington, 2005, pp. 15, 22 and 24.

In every year, side impacts ranked second only to frontal impacts as a cause of occupant fatalities in passenger cars. Technologies such as safety belts and frontal air bags are more effective in preventing fatalities in rollovers and/or frontals than in side impacts. Thus, deaths in lateral impacts, while shrinking in absolute numbers, now account for a larger share of the fatalities.

Nearside vs. farside: Occupants are especially at risk if they are sitting on the side of the car that was struck: drivers in left-side impacts and right-front passengers in right-side impacts. For these **nearside** occupants, only a car's relatively narrow side structure, comprising the doors, sill, roof rail and supporting pillars stands between the occupant and the impacting vehicle or object. That contrasts with frontal, rear and **farside** impacts where there is initially considerable distance and structure between the occupant and the contact. Figure 1-3 shows the ratio of nearside to farside fatalities is close to 2:1, year after year. Especially after 1990, close to 70 percent of the fatalities are nearside occupants, and just over 60 percent are front-outboard occupants (drivers and right-front passengers) in nearside impacts.

Figure 1-3: Nearside vs. Farside Fatalities, Passenger Cars, 1975-2004

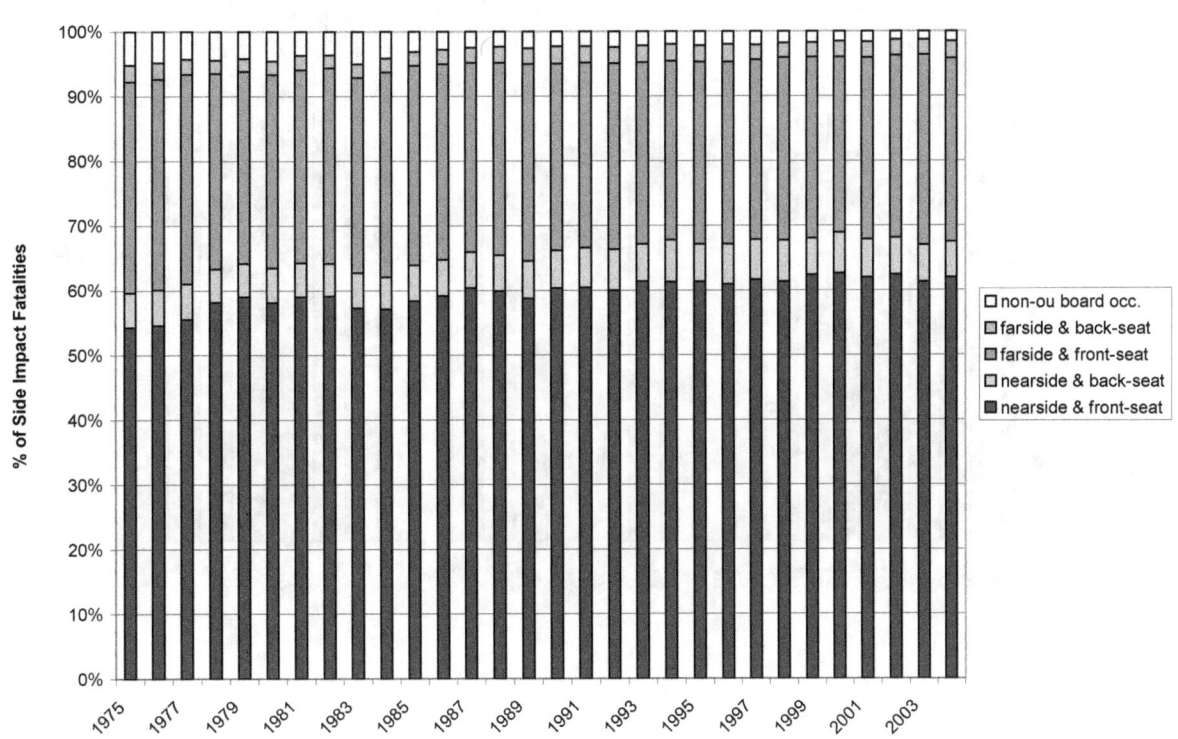

Multi- vs. single-vehicle crashes: Side impacts can occur when vehicles from two different roads collide front-to-side at an intersection, or when one vehicle, while turning or changing lanes, crosses the path of another vehicle on the same road. Occasionally, a vehicle can spin out of control and slide sideways into the path of another moving vehicle. The side of a car can impact a fixed object such as a tree or pole if the car runs off the road and spins out of directional control, sliding side-first into the object. The roadway departure may precede the loss of directional control, or vice-versa. Figure 1-4 indicates throughout 1975-2004 that close to 70 percent of the nearside, front-seat fatalities in passenger cars occurred in multivehicle crashes (involving 2 vehicles, or in some cases 3 or more vehicles), with perhaps some downward trend since 1997:

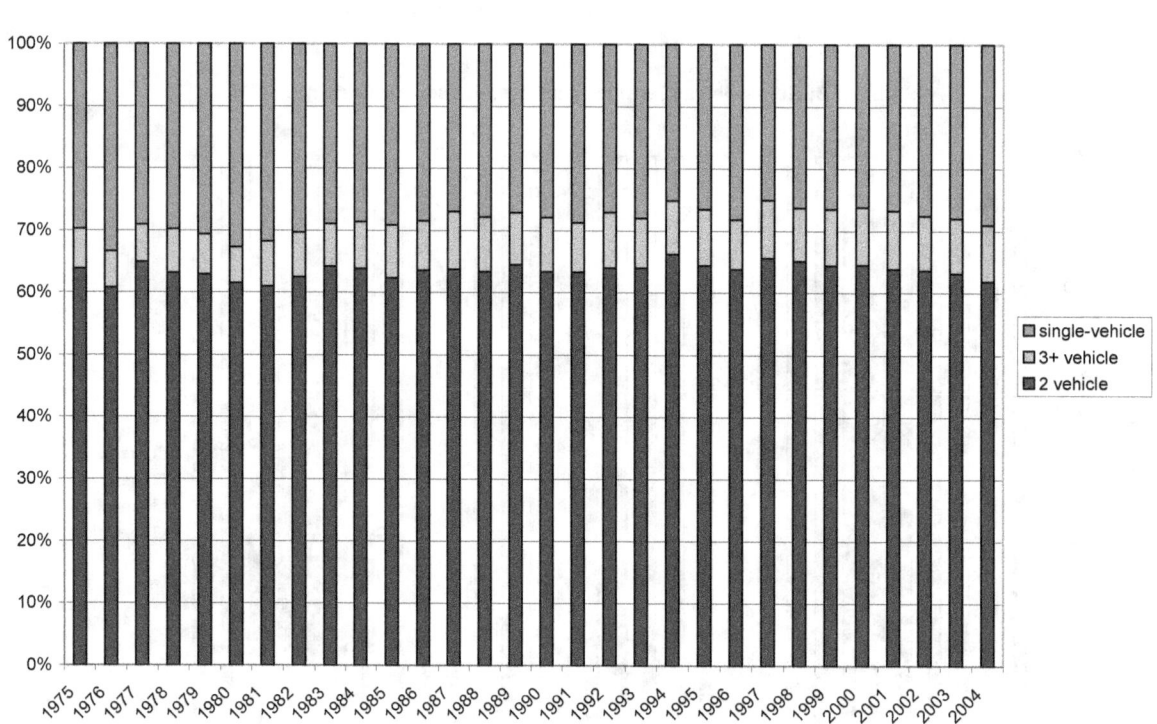

Figure 1-4: Percent of Nearside Fatalities that Are in Multi-Vehicle Crashes
Car Front-Seat Occupants, 1975-2004

Nearside front-seat fatalities in multivehicle crashes: Figure 1-5 shows that 3,000-3,500 drivers and right-front passengers of cars died each year in nearside impacts by other vehicles. As will be discussed later, they are a primary target population for the improvements envisioned in the 1990 amendment to FMVSS 214. Fatalities declined from an average of 3,500 in 1995-1997 to about 3,000 in 2004.

Figure 1-5: Nearside Fatalities in Multi-Vehicle Crashes
Front-Seat Occupants of Passenger Cars, 1975-2004

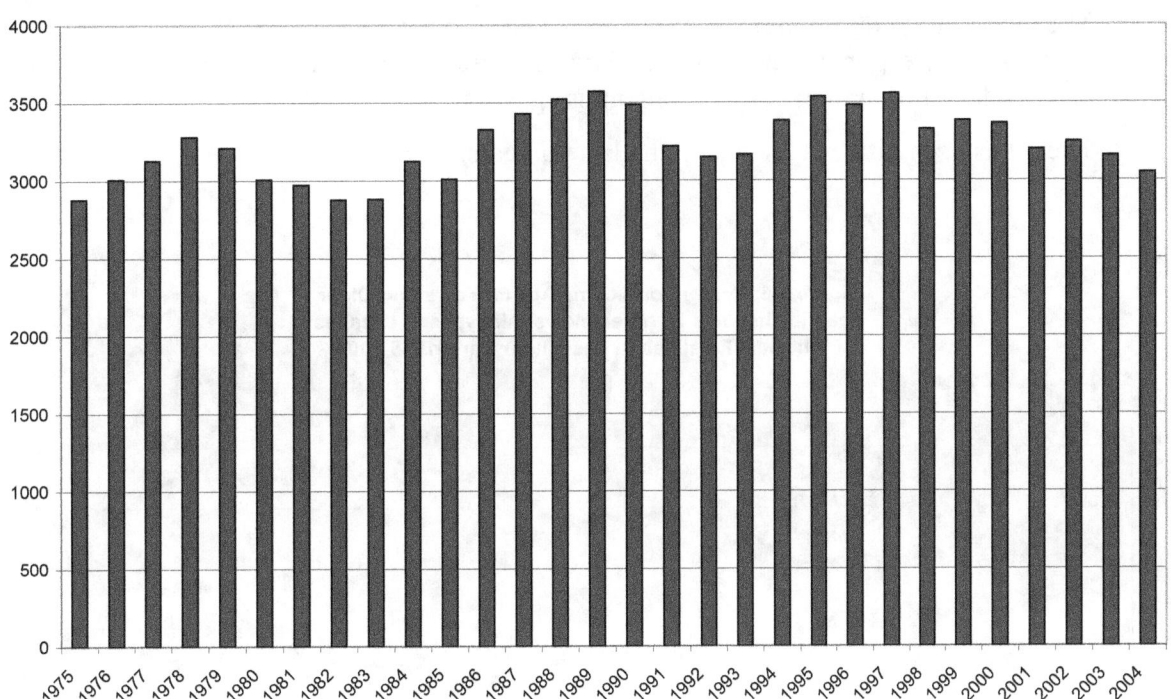

Older occupants: Figure 1-6 shows that a large proportion of the drivers and right-front passengers of cars killed in nearside impacts by another vehicle are 55 years or older, ranging from 33 percent in 1975 to 40-45 percent throughout 1991-2004 (blue line). By contrast, in all types of crash involvements of cars, including frontals and rollovers, only 19 percent of all driver and right-front passenger fatalities in 1975 and 25-30 percent in 1991-2004 are 55 years or older (red line). Older occupants are over-represented in the side impacts primarily because older drivers have more difficulty recognizing when it is safe to turn across oncoming traffic or enter an intersection. It is also conceivable that older occupants are especially susceptible to injury in this type of impact. Nevertheless, Figure 1-6 demonstrates that:

- The increasing proportion of older victims in nearside impacts since 1975 almost exactly parallels a corresponding increase in all types of crashes, both a consequence of an aging driver population (rather than a problem unique to nearside impacts).

- There has been little net change in the last 15 years.

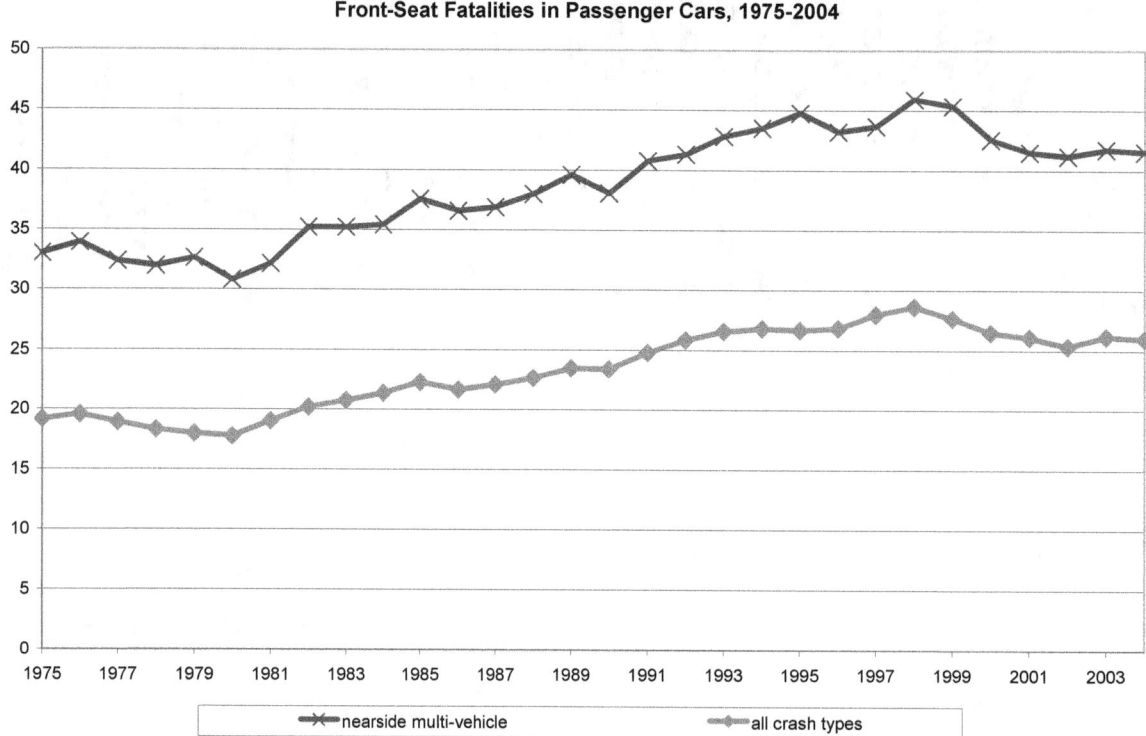

Figure 1-6: Percent of Victims Age 55 Years and Older
Nearside Impacts by a Vehicle vs. All Types of Crashes
Front-Seat Fatalities in Passenger Cars, 1975-2004

Striking vehicle type: One characteristic of nearside impacts did change dramatically during 1975-2004. Figure 1-7 shows that, in 1975, over 60 percent of the nearside fatalities to car occupants in two-vehicle crashes involved an impact by another passenger car and less than 20 percent were impacts by LTVs. By 2004, less than 30 percent were impacts by passenger cars and over 50 percent were impacts by LTVs. Impacts by heavy trucks accounted for 20 percent of the fatalities throughout 1975-2004. The shift to LTVs, of course, reflects the growing percentage of LTVs in the on-road fleet. Nevertheless, LTVs are over-represented as striking vehicles in fatal crashes relative to their share of registrations. In 2004, for example, there were 89,938,581 LTVs on the road and 133,275,377 passenger cars, yet LTVs outnumbered cars as the striking vehicle by more than 5 to 3.[4] Factors such as greater mass, height and rigidity make LTVs more aggressive than cars as a "bullet" vehicle in side impacts.[5]

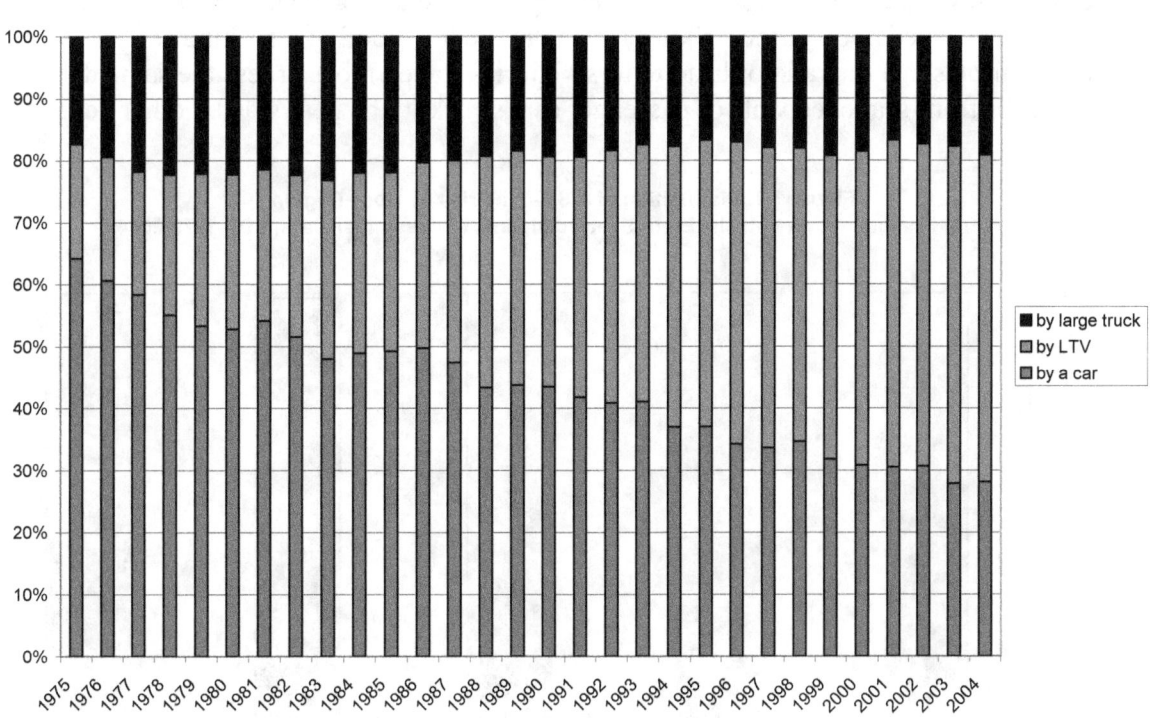

Figure 1-7: Striking Vehicle Type in 2-Vehicle Crashes
Nearside Fatalities to Front-Seat Occupants of Cars, 1975-2004

---

[4] *Ibid.*, pp. 22 and 24.
[5] Kahane, C.J., *Vehicle Weight, Fatality Risk and Crash Compatibility of Model Year 1991-99 Passenger Cars and Light Trucks,* NHTSA Technical Report No. DOT HS 809 662, Washington, 2003, Chapter 6; Gabler, H.C. and Hollowell, W.T., "NHTSA's Vehicle Aggressivity and Compatibility Research Program," Paper No. 98-S3-O-01, *Proceedings of the 16th International Technical Conference on the Enhanced Safety of Vehicles*, Report No. DOT HS 808 759, Washington, 1998; Gabler, H.C. and Hollowell, W.T., *The Aggressivity of Light Trucks and Vans in Traffic Crashes*, Paper No. 980908, Society of Automotive Engineers, Warrendale, PA, 1998.

Injury distribution by body region: During a side impact by another vehicle, the car's side structure has limited capacity to absorb energy. The structure is deflected into the passenger compartment nearly at the impact speed of the "bullet" vehicle and soon makes contact with the nearside occupant, especially the occupant's torso, because it tends to be on the same level as the striking vehicle's front. The Crashworthiness Data System (CDS) of the National Automotive Sampling System (NASS) documents the injuries in various types of crashes during 1979-2004. In frontal impacts (a benchmark), 56 percent of the life-threatening injuries – levels 4-6 on the Abbreviated Injury Scale (AIS) – to drivers and right-front passengers of passenger cars were to the occupant's torso and 44 percent to the head or neck. But in nearside impacts by another vehicle, 63 percent of life-threatening lesions are torso injuries. By contrast, in nearside impacts with fixed objects, which may contact the car from floor to ceiling, 50 percent were torso injuries. And in farside impacts, where occupants are in less danger of immediate contact with intruding structures but may be tossed around the vehicle, only 46 percent were torso injuries.

The shift from passenger cars to LTVs as the predominant striking vehicle raises the question that head injuries could have increased substantially because the occupant's head is more likely to contact the elevated hood of the striking LTV than the low hood of a striking car.[6] Figure 1-8, however, indicates that the ratio of head to torso injuries stayed more or less the same throughout 1979-2004 for front-seat occupants of passenger cars struck in the near side by other vehicles:

Figure 1-8: Distribution of AIS 4-6 Injuries by Body Region
Nearside Impacts by a Vehicle, Front-Seat Occupants of Passenger Cars, CDS 1979-2004

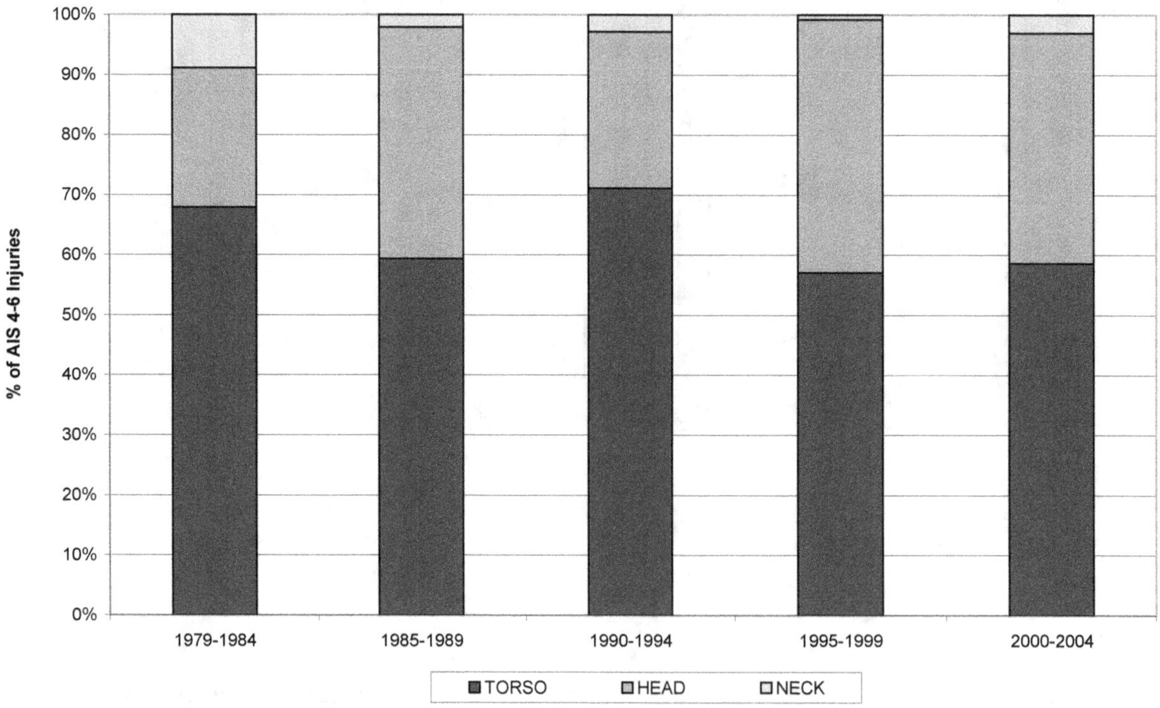

---

[6] Dalmotas specifically asked it in his peer review.

Figure 1-8 analyzes the distribution of individual AIS 4-6 **injuries**, possibly more than one per occupant. Figure 1-9 performs the same analysis at the **person** level. A person who has one or more AIS 4-6 injuries may have such injuries to the torso only, the head/neck only or possibly to both. Figure 1-9 shows that, throughout 1979-2004, close to 55 percent of the front-outboard occupants with life-threatening injuries in nearside impacts by other vehicles had such injuries to the torso only, and close to 76 percent had a torso injury plus, possibly a head or neck injury. Just 24 percent had head or neck injuries alone. In frontal crashes, 38 percent had only head or neck injuries; in nearside impacts with fixed objects and in all farside impacts, 48 percent.

Figure 1-9: Distribution of Occupants by Body Regions with AIS 4-6 Injuries
Nearside Impacts by a Vehicle, Front-Seat Occupants of Passenger Cars, CDS 1979-2004

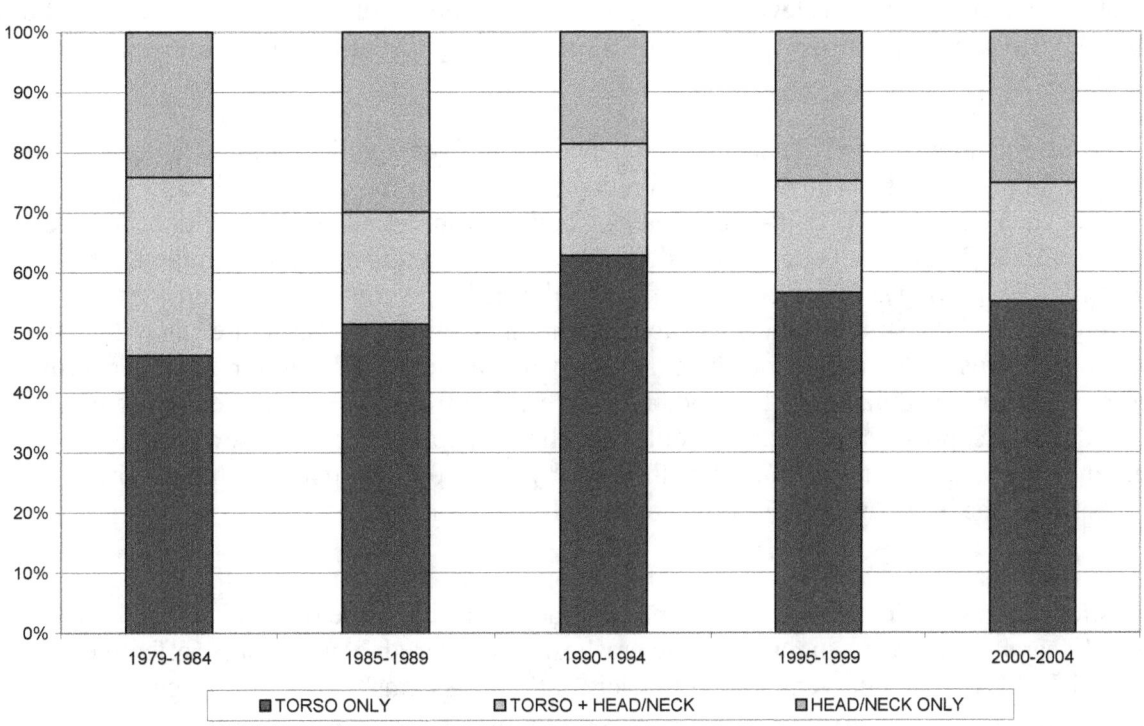

Belt effectiveness: Even though safety belts are quite effective in almost every other type of crash, they are of little help in a nearside impact, directly into the occupant compartment, by another vehicle. Torso contact with the intruding side structure is likely to occur whether an occupant is belted or not. NHTSA found only a non-significant 5 percent reduction in fatality risk for belt use in nearside multivehicle crashes, as compared to statistically significant reductions of 21 percent in nearside impacts with fixed objects, 39 percent in farside impacts, 50 percent in frontals, and 74 percent in rollovers.[7]

---

[7] Kahane, C.J., *Fatality Reduction by Safety Belts for Front-Seat Occupants of Cars and Light Trucks*, NHTSA Technical Report No. DOT HS 809 199, Washington, 2000, pp. 28-32; specifically, pp. 31-32 explain that safety belts are more effective in nearside impacts with fixed objects than in nearside impacts by other vehicles primarily because the former are far more likely to involve occupant ejection.

1.2 Side door beams: an early measure to protect occupants

Before 1969, the side doors of passenger cars were nearly empty shells of sheet metal, offering little protection to occupants in side impacts. **Side door beams**, running longitudinally inside the door, were a first step to provide some crush resistance and structural strength. During the 1960's, Hedeen and Campbell at General Motors developed the beams and a static test for measuring a door's crush resistance. They were installed in MY 1969 full-size GM cars. By then, NHTSA had announced its intention to regulate side door strength with an Advance Notice of Proposed Rulemaking (ANPRM) in October 1968. The first version of Federal Motor Vehicle Safety Standard (FMVSS) 214 was issued as a Final Rule in October 1970, with an effective date of January 1, 1973. It sets strength requirements for side doors, based on a static test of crush resistance: a rigid steel cylinder is gradually forced into the door, and it must encounter crush resistance exceeding various levels that depend on the depth of the crush and the weight of the car. All cars were equipped with side door beams meeting FMVSS 214 at some point during MY 1969-1973.[8]

A typical side door beam is a metal bar of channel (fluted) design, 8 inches wide, located inside the door, about 10 inches above the sill, running the length of the door, attached to the door frame at each end. In MY 1979-1981, beams weighed from 5 to 7 pounds per door in 4-door cars, and 10 to 21 pounds per door in 2-door cars. Any structure added within the door is welcome. By putting some crush on the front of the striking vehicle and/or transmitting force to the remainder of the struck vehicle and accelerating it sideways, the structure can reduce the amount of intrusion toward the occupant and slow down the rate of that intrusion. Nevertheless, researchers suspected that a 5-21 pound beam, stretching from one end of the door to the other without much support in the middle, would have limited power to resist a severe and perpendicular impact into the middle of the door by a 2,000-5,000 pound vehicle. The 10-inch gap between the beam and the sill is an additional weak point.

NHTSA evaluated the fatality and injury reduction of side door beams in 1982, based on statistical analyses of crash data.[9] As expected, the beams had little or no effect on the fatality risk of nearside occupants in multivehicle crashes. However, they were rather effective in some other situations. In single-vehicle side impacts, fatality risk was reduced by 14 percent for nearside and farside occupants, and when this group of crashes was further limited to impacts with a single fixed object, fatality reduction was 23 percent. Here, rather than merely absorbing energy, the beam acts like an internal "guard rail" to allow a car to slide past a pole or tree, with a longer, shallower crush pattern on the car. Integrity of the side structure was better preserved. Beams were also effective in somewhat lower-speed multivehicle crashes, reducing the risk of nonfatal injuries. When the damage was centered in the occupant compartment area, side door beams reduced nearside occupants' hospitalizations by a statistically significant 25 percent. NHTSA estimates that side door beams saved nearly 500 lives in single-vehicle crashes of

---

[8] Kahane, C.J., *Lives Saved by the Federal Motor Vehicle Safety Standards and Other Vehicle Safety Technologies, 1960-2002,* NHTSA Technical Report No. DOT HS 809 833, Washington, 2004, pp. 136-140; Kahane, C.J., *An Evaluation of Side Structure Improvements in Response to Federal Motor Vehicle Safety Standard 214,* NHTSA Technical Report No. DOT HS 806 314, Washington, 1982, pp. 100-108; Hedeen, C.E. and Campbell, D.D., *Side Impact Structures,* Paper No. 690003, Society of Automotive Engineers, New York, 1969; *Federal Register* 33 (October 5, 1968): 14971, 35 (October 30, 1970): 16801.
[9] Kahane (1982).

passenger cars in 2002, and have also prevented about 9,500 nonfatal hospitalizations per year in single- and multivehicle crashes.[10]

NHTSA extended the static strength test of FMVSS 214 to LTVs, effective September 1, 1993. In single-vehicle side impacts, side door beams reduced fatality risk in LTVs by a statistically significant 19 percent. NHTSA's evaluation estimated that side door beams would eventually save 151 lives per year in LTVs, when all LTVs on the road have the beams.[11]

1.3 The dynamic test requirement for FMVSS 214

By the late 1970's, if not earlier, researchers suspected that side door beams alone would not sufficiently attenuate intrusion in a severe side impact by another vehicle to reduce fatality risk to the nearside occupant of the struck car. At a public Side Impact Conference on January 31, 1980, NHTSA outlined its plans to upgrade FMVSS 214 with a dynamic test.[12] The new regulation aimed to reduce fatality risk to the nearside occupant when a car is struck in the door area by another vehicle - the configuration responsible for the largest group of side impact fatalities – and especially to reduce fatal thoracic injuries.

Unlike some earlier FMVSS that could draw upon extensive information about existing test procedures and safety technologies, the FMVSS 214 upgrade necessitated many years of research, analysis and testing by NHTSA and others in the safety community. Researchers from the United States and other countries considered numerous alternative injury criteria, dummies, test configurations, etc. NHTSA's selected approach comprised:

- A review of crash data, indicating that the archetypal side impact fatality in the 1980's involved a fast-moving car striking a slow-moving car in the door, at a right angle: a typical intersection collision.

- A review of injury data, indicating that a large proportion of the nearside occupants' life-threatening injuries occurred when the sides of their torsos contacted the interior side surface (most frequently the door) of the car. (Head injuries, as noted above, are also a frequent cause of fatalities in side impacts, but were not the principal focus of this rulemaking process. Recent and ongoing rulemaking to address head injuries are discussed in Section 1.4.)

- The Thoracic Trauma Index (TTI) was found to be an excellent predictor of thoracic injury severity in experimental side impacts to cadavers.[13] $TTI = \frac{1}{2} (G_R + G_{LS})$, where G_R is the greater of the peak accelerations of either the upper or the lower rib, expressed in g's and G_{LS} is the lower spine (T12 vertebra) peak acceleration. Pelvic g's are an additional injury criterion, but TTI is the key predictor of life-threatening injuries.

---

[10] Kahane (2004), pp. 140 and 217.
[11] Walz, M.C., *Evaluation of FMVSS 214 Side Impact Protection for Light Trucks: Crush Resistance Requirements for Side Doors*, NHTSA Technical Report No. DOT HS 809 719, Washington, 2004; *Federal Register* 56 (June 14, 1991): 27427.
[12] *Side Impact Conference*, NHTSA Report No. DOT HS 805 614, Washington, 1980.
[13] *Final Regulatory Impact Analysis - New Requirements for Passenger Cars to Meet a Dynamic Side Impact Test FMVSS 214*, NHTSA Publication No. DOT HS 807 641, Washington, 1990.

- Development of a Side Impact Dummy (SID) on which TTI (as well as pelvic g's) can be reliably measured in a side impact test configuration. The injury score measured on the dummy is called TTI(d).

- A Moving Deformable Barrier (MDB) was developed to represent a generic 3000-pound passenger vehicle. The test procedure simulates an MDB moving 30 mph hitting, at a right angle, the door area of a subject vehicle, traveling 15 mph. (It is accomplished by having the MDB travel at 33.54 mph at an angle of 63 degrees with respect to the longitudinal centerline of a stationary test vehicle. The wheels of the MDB are "crabbed" 27 degrees toward the rear of the test vehicle to obtain a right-angle contact.)

- Testing various production 1980-1988 passenger cars to learn the baseline distribution of TTI(d). Some baseline testing continued after the Final Rule was issued in 1990, up to model year 1993, just before the phase-in period for the MDB test requirement.

- Demonstration of two technologies, **structure** and **padding**, that, singly or in combination can significantly improve (i.e., decrease) TTI(d) from its baseline levels in production vehicles.

- Regulatory analysis[14] to estimate the lives saved by decreasing TTI(d) to various levels, and the extent of vehicle modifications needed to secure those levels – and, finally –

- On October 30, 1990, NHTSA issued the Final Rule amending FMVSS 214 to phase in a dynamic test of side impact protection during model years 1994-1997. FMVSS 214 recognizes the greater difficulty of protecting occupants in 2-door cars. FMVSS 214 allows TTI(d) up to 90 in 2-door cars, but limits 4-door cars to 85. FMVSS 214 also includes test limits on pelvic g's and has door retention requirements to reduce occupant ejection. At least 10 percent of passenger cars produced between September 1, 1993 and August 31, 1994 had to meet the standard; at least 25 percent of cars produced between September 1, 1994 and August 31, 1995; at least 40 percent of cars between September 1, 1995 and August 31, 1996; and all cars after September 1, 1996. During that phase-in period, manufacturers declared ("self-certified") what make-models complied with FMVSS 214. NHTSA advised the public on what models were certified.[15]

- The regulatory analysis projected that at least 512 lives would be saved per year if TTI(d) improved from its baseline levels in cars of the mid-1980's to 90 or better in all 2-door cars and 85 or better in all 4-door cars.[16]

The new version of FMVSS 214, however, retained the original "static" test in view of the demonstrated effectiveness of side door beams in collisions with fixed objects. Furthermore, the side door beam, often strengthened, continued to be an integral part of the structures used to meet the dynamic test requirement.

---

[14] *Ibid.*

[15] *Federal Register* 55 (October 30, 1990): 45752; *NHTSA Hails Safety Features in Model Year 1994 Passenger Cars and Light Trucks and Vans*, Press Release No. NHTSA 38-93, U. S. Department of Transportation, Office of the Assistant Secretary for Public Affairs, Washington, 1993.

[16] *Final Regulatory Impact Analysis,* p. IV-62; includes 498 lives saved by mitigating thoracic injuries plus 14 lives saved by preventing occupant ejection through better door retention; a deduction was made for projected increases in safety belt use.

Side Impact NCAP In addition to compliance tests that assure cars meet the minimum requirements of FMVSS 214, NHTSA provides consumer information on vehicle performance in side impacts. The information is collected as part of NHTSA's New Car Assessment Program (NCAP) and posted on the agency's web site, www.safercar.gov . The agency uses a rating system of one star (worst) to five stars (best) for front-outboard and rear-outboard occupants, based primarily on TTI(d) but also taking into account pelvic g's. In the NCAP tests, the MDB strikes the side of the target vehicle at 38.5 mph, 5 mph faster than in the FMVSS 214 test. The purpose of the higher speed is to differentiate more clearly between average and superior performance in severe crashes. The side NCAP program started shortly after September 1, 1996, the date when all new cars were required to meet the dynamic side impact test of FMVSS 214. Side NCAP has provided an additional incentive to decrease TTI(d) well below the requirements set by FMVSS 214 and to seek further improvements in TTI(d) as time goes on.

European regulations and NCAP The European Union approved a side impact safety regulation, EU Directive 96/27/EC, in October 1996. It applies to all new or redesigned models manufactured after October 1, 1998, and all other vehicles manufactured after October 1, 2003. Like FMVSS 214, a MDB is launched into a stationary target vehicle occupied by one dummy in the front seat. However, the test speed is slightly lower (50 kph) and there is no crab angle – i.e., no attempt is made at simulating the movement of the target vehicle. The MDB is lighter (2,095 lbs), smaller and softer than in FMVSS 214, although 0.8 inches higher off the ground. As in FMVSS 214, successful test performance is determined by dummy injury criteria. However, both the test dummy and injury criteria differ from those in FMVSS 214. SID is capable of measuring acceleration of the ribs, spine and pelvis. A dummy called Euro SID is used instead of SID. It measures force and displacement as well as acceleration-based readings. The regulation limits the Head Injury Criterion (HIC) to 1000, rib deflection to 42 mm (1.7 in.), the Viscous Criterion (V*C) to 1 m/s, abdominal force to 2.5 kN (562 lbs) and the force on the pubic symphysis region to 6 kN (1350 lbs).[17]

The Euro NCAP program began side impact testing in 1996 and published its first results in 1997. Unlike the United States, the NCAP test speed is the same as the EU regulation (50 kph). Initially, there were four star ratings. In 2000, Euro NCAP added a voluntary pole test that can improve the side impact score and potentially add a fifth star to the rating for side impact. In 2003, the more advanced Euro SID 2 superseded the Euro SID dummy.[18]

Side impact ratings by the Insurance Institute for Highway Safety (IIHS) began in 2003. Their MDB weighs 3,300 pounds (300 pounds more than FMVSS 214) and its front end simulates the height and other characteristics of a pickup truck or SUV. The test speed is slightly

---

[17] Anders Lie recommended a discussion of European regulations and Euro NCAP in his review of this report. *NHTSA Plan for Achieving Harmonization of the U.S. and European Side Impact Standards, Report to Congress, April 1997,* NHTSA Docket No. NHTSA-1998-3935-1, 1998. The Viscous Criterion is calculated from combined rib displacement and velocity.

[18] *Creating a Market for Safety – 10 Years of Euro NCAP,* European New Car Assessment Programme, Brussels, 2005, accessible from www.euroncap.com ; McNeill, A., Haberl, J., Holzner, M., Schoeneburg, R., Strutz, T. and Tautenhahn, U., "Current Worldwide Side Impact Activities – Divergence versus Harmonisation and the Possible Effect on Future Car Design," Paper No. 05-0077, *Proceedings 19th International Technical Conference on the Enhanced Safety of Vehicles,* NHTSA Report No. DOT HS 809 825, Washington, 2005, accessible from www-nrd.nhtsa.dot.gov/departments/nrd-01/esv/19th/esv19 htm .

lower (50 kph) and there is no crab angle. The dummies are SID-II 5th percentile females, considered at greater risk than the 50th percentile male. IIHS rates vehicles good, acceptable, marginal or poor, based on: injury criteria for the head/neck, torso and pelvis/leg; movements of the dummy's head; and the vehicle's structural performance.[19]

1.4 Technologies to protect occupants in side impacts

Circa 1990, NHTSA believed that manufacturers would be able to meet Standard 214 (with just passing scores) by installing only padding in many cars. Some cars might need structural modifications, especially 2-door cars. Other cars might not need any change at all, especially the larger 4-door cars. Manufacturers might modify structure more extensively if they aimed to drop TTI(d) well below the FMVSS 214 requirement. By the mid-1990's, the industry was already well on its way to developing air bags that deploy and offer additional protection in side impacts.

Padding reduces the probability of occupant injury, given that the door structure has contacted the occupant. The padding is located within the door at points where hip or chest contacts are likely. It is thick plastic foam - not a soft pad – capable of absorbing significant energy at a force-deflection rate safe for occupants. Without the padding, more rigid components would immediately contact the occupant.

Structure modifications, beyond the side door beams installed in response to the original, static test, slow down and reduce the extent of door intrusion into the passenger compartment. They included substantially strengthening the beams themselves and/or the pillars, sills, roof rails, seats or cross-members of a car, and strengthening the overlap between doors and pillars, sills, etc. The test procedure enables manufacturers to identify the weakest points in the structure of their prototype cars and reinforce them as needed.

Side-impact air bags

Torso air bags During the 1990's, manufacturers and suppliers developed air bags that deploy from the seat or the door to provide an energy-absorbing cushion between the occupant's torso and the vehicle's side structure during lateral impacts. Conceptually, torso air bags do the same thing as padding, but they do a lot more of it. Volvo made them standard on all their MY 1996 cars, while Audi, BMW and Cadillac began to furnish them as standard equipment on some 1997 models and offer them as options on others. By MY 2001, nearly 30 percent of new cars were equipped with torso air bags, and that percentage stayed about the same in 2002 and 2003. They can substantially improve TTI(d), as we shall see in Section 1.5. NHTSA's annual *Buying a Safer Car* brochures inform the public what make-models are equipped with torso air bags.[20]

Head-protection air bags Measures to decrease TTI(d) are first and foremost designed to mitigate torso injuries, although they may also reduce head injuries. Head-protection air bags, on the other hand, specifically target head injuries, which account for 37-54 percent of life-

---

[19] www.iihs.org/ratings .

[20] *Buying a Safer Car 2000*, NHTSA Publication No. DOT HS 809 046, Washington, 2000; *Buying a Safer Car 2001*, NHTSA Publication No. DOT HS 809 152, Washington, 2000; *Buying a Safer Car 2002*, NHTSA Publication No. DOT HS 809 409, Washington, 2002; *Buying a Safer Car*, NHTSA Publication No. DOT HS 809 546, Annual publication, 2003-2005.

threatening lesions in various types of side impacts (Section 1.1). They may have an additional benefit as a barrier to occupant ejection through side windows. By the mid-1990's, auto industry suppliers were developing head-protection airbags for meeting the proposed FMVSS 201. On July 29, 1998, NHTSA amended FMVSS 201 (occupant protection in interior impact) to facilitate the introduction of these air bags.[21] BMW introduced head air bags as standard equipment in some lines in model year 1998, and by 2001, many of the large manufacturers offered them as standard or optional equipment on various models. By MY 2003, nearly 20 percent of new passenger cars were equipped with some type of head air bag. On May 17, 2004, the agency issued a Notice of Proposed Rulemaking (NPRM) to amend FMVSS 214, proposing to add a 20 mph side impact with a pole, at a 75-degree angle (i.e., 15 degrees forward of a purely lateral impact). The proposed three-year phase-in dates would start four years after publication of a Final Rule. NHTSA anticipates that head air bags would generally be installed to meet the new requirement.[22]

There are currently two distinct types of head-protection air bags:

- "Curtains" or "tubes (inflatable tubular structures)" that drop down from the roof rail into the side-window area. These are separate from any torso air bags in the vehicle, although they usually share components such as sensors and the control module. Initially, all vehicles equipped with head curtains or tubes also had torso air bags, but starting in 2001, some vehicles were equipped with head curtains only, and no torso air bags.

- "Torso/head combination bags" that deploy from the seat to protect the torso but also extend upward far enough to protect the head impact zones around the side window.

NHTSA's annual *Buying a Safer Car* brochures inform the public what make-models are equipped with head-protection air bags, and the type of bags.[23]

The head injury protection upgrade for FMVSS 201

On August 14, 1995, NHTSA issued a Final Rule extending the head injury protection requirements of FMVSS 201. It established a new list of target areas in the vehicle's upper interior, including the A-, B- and other pillars, the front and rear roof header, the roof side rails, and the upper roof, among others. It is not a side impact standard *per se*, because these structures can be sources of life-threatening head injuries in any crash mode, and they are located on the front, rear and top as well as the sides of the vehicle. Nevertheless, side impacts account for many of the injuries. In a 15 mph impact test of a free-motion headform (FMH) into any of these targets, the Head Injury Criterion (HIC) may not exceed 1000 for any 36-millisecond

---

[21] *Federal Register* 63 (August 4, 1998): 41451; Recognizing that the 15 mph headform test might be a problem in target areas where the undeployed air bag is stored (and, furthermore, an inappropriate test if the bag usually deploys at that speed), NHTSA offered an alternative compliance procedure. Manufacturers have the option to reduce the speed of the headform test to 12 mph on target areas where the bag is stored, provided they can meet an 18 mph lateral (90 degree) crash test for the full vehicle into a pole – with HIC < 1000. The pole test simulates a side impact with a fixed object (e.g., a tree, utility pole or concrete abutment) and it measures the severity of the head impact with the deployed bag.
[22] *Federal Register* 68 (May 17, 2004): 27990.
[23] *Buying a Safer Car 2000-2005.*

period. Impacts may be directed from a range of vertical and horizontal angles, not just head-on.[24]

The evaluation of FMVSS 201 is a high priority for NHTSA, but outside the scope of this report, because many years of detailed data on injuries by body region and injury source will be needed.[25] In this report, we are concerned with FMVSS 201 primarily to the extent that it could interact, or be a confounding factor in our evaluations of TTI(d) improvements, torso bags and head air bags.

Manufacturers were offered a choice of several alternative phase-in schedules during the four years from September 1, 1998 to September 1, 2002. For example, they could certify the new requirements on at least 10 percent of cars and LTVs manufactured during the first year, at least 25 percent during the second year, at least 40 percent during the third year, at least 70 percent during the fourth year, and all cars and LTVs manufactured on or after September 1, 2002.

Manufacturers could certify to FMVSS 201 by:

- Adding energy-absorbing materials such as padding, ribbing, or an "egg-crate" honeycomb configuration around target areas, or using a thicker roof liner.

- Adding head air bags; in fact, as mentioned above, NHTSA's 1998 amendment of FMVSS 201 facilitated the use of air bags.

- A combination of both, relying on energy-absorbing materials in target areas not covered by the air bag.

- Little or no change, if a pre-standard vehicle could already meet FMVSS 201 at most or all target areas.

Given those alternatives, it is not surprising that the phase-in period for FMVSS 201 overlapped the initial installations of head air bags. Many make-models were certified to FMVSS 201 with energy-absorbing materials one or more years before they offered head air bags – i.e., head impact protection was upgraded in two distinct, temporally separate stages. But many others certified at the same time or even after they offered them – usually, but not always signifying that the entire upgrade, air bags plus energy-absorbing materials (if any) was implemented at once.[26] But FMVSS 201 certification (without head air bags) also overlapped the initial installation of torso bags only in quite a few make-models and sometimes even coincided with "second generation" TTI(d) improvements to structures and padding that took place after the initial 1994-1997 phase-in of FMVSS 214.

---

[24] *Federal Register* 60 (August 18, 1995): 43031; Kahane (2004), p. 51.

[25] *National Highway Traffic Safety Administration Evaluation Program Plan, Calendar Years 2004-2007*, NHTSA Report No. DOT HS 809 699, Washington, 2004, p. 8.

[26] Possible reasons for not certifying a make-model to FMVSS 201 until a year or more after offering head air bags could include: (1) head air bags were optional, not furnished on every vehicle; (2) the manufacturer, being ahead of the phase-in schedule, had no obligation to certify this make-model, even though it would have complied with FMVSS 201; (3) the manufacturer had to make additional changes in subsequent years before the vehicle met FMVSS 201.

NHTSA's cost analyses of FMVSS 201 comprise 15 make-models including 14 that had offered standard or optional head air bags by 2004.[27] Of these 14, seven certified to FMVSS 201 with energy-absorbing materials one or more years before they offered head air bags – i.e., they upgraded head impact protection in two separate stages, whereas the other seven certified at the same time or even after they offered head air bags. The analyses identified tangible and relatively substantial additions of energy-absorbing material in five of the seven models that initially certified to FMVSS 201 without head air bags, whereas two certified with minor modification. But when these seven models were subsequently upgraded with head air bags, there were few additional changes in the energy-absorbing materials. Likewise, in the other seven models that initially certified to FMVSS 201 with head air bags, the energy-absorbing materials also changed little at that time.[28] In both groups, the installation of head air bags was generally not accompanied by a substantial upgrade (or downgrade) in the energy-absorbing materials that provide head impact protection.

1.5   What actually happened: average TTI(d), 1981-2002

Background A "typical" NHTSA rulemaking process creates a new performance requirement that is fulfilled in all cars by adding more or less the same specific equipment. All cars have it by the effective date, perhaps a year or two earlier in some cars but in any case after the rulemaking process is underway. The equipment was nonexistent or rare before the rulemaking process started, and it did not change in any important way in the years after the effective date. Center High Mounted Stop Lamps, installed in all 1986 cars and some 1985's, and little changed since then, are a good example of the typical process.[29] In short, we know what happened – and essentially the same thing happened on every make-model – and we know when it happened on each make-model. Evaluation is a relatively straightforward matter of comparing the crashes of vehicles before and after the equipment was installed.

Side structure improvements differ from the typical process in several important respects:

- The dynamic test requirement of FMVSS 214 did not result in the fleet-wide installation of any specific piece of equipment. Different components were modified, depending on the make-model. Furthermore, specific modifications can be difficult to identify if initial FMVSS 214 compliance was "built in" as part of an "integrated platform redesign" of that model.

- Side impact performance is measured by the continuous variables, TTI(d) and pelvic g's. TTI(d) has ranged from 32 to 131 on individual test vehicles. Whereas any TTI(d) up to 85 (90 in a 2-door car) is a "pass" and anything above that is a "fail," there are large differences of performance within the "pass" group and within the "fail" group.

---

[27] Ludtke, N.F., Osen, W., Gladstone, R. and Lieberman, W., *Perform Cost and Weight Analysis, Non Air Bag Head Protection Systems, FMVSS 201*, NHTSA Technical Report No. DOT HS 809 810, Washington, 2003; Ludtke, N.F., Osen, W., Gladstone, R. and Lieberman, W, *Perform Cost and Weight Analysis, Head Protection Air Bag Systems, FMVSS 201*, NHTSA Technical Report No. DOT HS 809 842, Washington, 2004.

[28] Ludtke et al. (2004), pp. 3-47 – 3-54; i.e., there were, on the whole, no substantial cost increases (or decreases) in the components that house the energy-absorbing materials.

[29] Kahane, C.J. and Hertz, E., *The Long-Term Effectiveness of Center High Mounted Stop Lamps in Passenger Cars and Light Trucks*, NHTSA Technical Report No. DOT HS 808 696, Washington, 1998.

- TTI(d) began to improve in some make-models well before the FMVSS 214 phase-in. The 1980-1993 development of FMVSS 214 was an iterative process with extensive public participation. Manufacturers could compare TTI(d) in their existing vehicles to levels proposed for FMVSS 214 or achieved by competitors, and correct their worst performers.

- Conversely, if models already had acceptable TTI(d) by MY 1993, manufacturers might certify their 1994 models as FMVSS 214-compliant despite changing nothing, or very little from the 1993's.

- Manufacturers could aim for much better TTI(d) than the 85/90 allowed by FMVSS 214 – upon initial certification, or in subsequent improvements. Side NCAP lets the public know when performance has improved. Subsequent improvements could be the result of adding or redesigning structure and padding, or installing torso and/or head air bags.

In other words, the evaluation of FMVSS 214 is not a simple comparison of two internally homogeneous groups, one "before" and one "after" certification. Side impact protection is a story of ongoing improvement(s), varying from model to model in magnitude and timing. Some models with acceptable performance before 1994 might not have improved much at all. To properly evaluate the effect of side impact protection on a model's fatality risk, we should know its TTI(d) history and identify in what years scores improved and by how much. We should also learn why the scores improved at that time: whether due to structure, padding and/or air bags. That will make it possible to identify groups of make-models that significantly improved side impact protection at specific times and compare their fatality risk before and after the change.

What actually happened *in 1994-1997* The manufacturers provided NHTSA with detailed lists and diagrams showing changes they made to achieve compliance during the phase-in period for FMVSS 214, model years 1994-1997. Structural modifications and padding were the principal technologies used to meet FMVSS 214 in those years. This information suggests that make-models accounting for approximately:

- 56 percent of new car sales received substantial structural modifications, usually accompanied with padding. "Substantial" structure could include extensive strengthening or reinforcement of side door beams; A-, B- or C- pillars; sills; roof rails; seat structures or cross-members of a car: typically 4 or 5 such items per car.

- 21 percent of cars received padding with minor structural modifications. "Minor" structure could include small, localized reinforcements on the components listed above, or even some extensive strengthening, but to at most one or two major components.

- 6 percent of cars received padding only.

- 17 percent remained essentially unchanged from previous years, implying that even the pre-1994 models of these cars could have met FMVSS 214.

- Less than 1 percent of 1994-1997 cars had side-impact air bags.[30]

Before 1994 As we shall see, TTI(d) began to improve overall during the 1980-1993 development of FMVSS 214, well before the actual 1994-1997 phase-in. However, the agency has little information to track and explain changes in TTI(d) performance in specific make-models. We can surmise that improvements were due primarily to structural modifications, simply because the energy-absorbing padding typically in the side doors of today's cars was not in wide use before 1994. We may also surmise that improvements were most extensive in 2-door cars, because they originally were the worst performers. For example, one manufacturer ran a cross member from the left to the right A-pillar through the dash of a 2-door model, reinforced the B-pillar at the sill level and added some floor stiffeners.[31]

After 1997 Although some manufacturers revised their side structures, or redesigned their models with improved side structures after 1997, the predominant change was the introduction of torso and/or head air bags. Typically the change came in two waves: initially torso-only air bags, followed by extending these bags upward, or providing separate curtains for head protection. In MY 2003, approximately

- 11 percent of new cars had torso air bags, but no air bags for head protection.

- 7 percent: torso/head combination bags.

- 8 percent: torso air bags and separate head curtains or tubes.

- 3 percent: head curtains only.[32]

Average TTI(d) of new cars in model years 1981-2002 NHTSA has test results for approximately 375 production vehicles of model years 1981-2002 that were impacted by a moving deformable barrier (MDB) in the FMVSS 214 test configuration, for which TTI(d) was successfully measured on the front-seat dummy.

These tests were conducted at various different speeds. Typically, compliance tests on post-standard cars are conducted at slightly below the nominal FMVSS 214 test speed of 33.54 mph, research tests on earlier vehicles at slightly above 33.54 mph, and side NCAP tests at just below 38.5 mph. An adjustment factor for the different test speeds, and a relationship between test speed and TTI(d) was identified by analyzing 31 pairs of cars: one specimen was compliance-tested and another specimen of the same make-model (or an essentially identical corporate cousin), the same model year, and the same type of side air bag (or lack thereof) was NCAP tested. For these 31 pairs, the average value of

---

[30] Kahane, C.J., *Evaluation of FMVSS 214 - Side Impact Protection: Dynamic Performance Requirement; Phase 1: Correlation of TTI(d) with Fatality Risk in Actual Side Impact Collisions of Model Year 1981-1993 Passenger Cars*, NHTSA Technical Report No. DOT HS 809 004, Washington, 1999, pp. vii and 139-143; Tarbet, M.J., *Cost and Weight Added by the Federal Motor Vehicle Safety Standards for Model Years 1968-2001 in Passenger Cars and Light Trucks*, NHTSA Technical Report No. DOT HS 809 834, Washington, 2004, pp. 115-119.
[31] Kahane (1999), pp. vii, 6 and 19-23.
[32] *Preliminary Economic Assessment, FMVSS 214, Amending Side Impact Dynamic Test Adding Oblique Pole Test*, NHTSA Docket No. NHTSA-2004-17694-1, 2004, p. VI-12.

$$\frac{\log(\text{ncap TTI}) - \log(\text{compliance TTI})}{\log(\text{ncap speed}) - \log(\text{compliance speed})}$$

was 1.89, and its standard error was 0.18. In other words, the empirical elasticity of TTI(d) to test speed was close to 2, and it did not differ significantly from 2. That would be an intuitively reasonable relationship, in that the acceleration of the dummy's torso ought to be proportional to its kinetic energy immediately after impact, and to the square of the impact speed. In the remainder of this analysis, the actual TTI(d) in the test is replaced by the adjusted value that would likely have been observed if the test had been run at exactly 33.54 mph.

$$\text{TTI(d) adjusted} = \text{TTI(d) observed} * (33.54/\text{SPEED})^2$$

A test car's TTI(d) applies not only to all cars of the same make-model, model year and side-air-bag status as the test vehicle, but also to other cars that are essentially identical:

- Earlier and later model years of the same make-model if there was no redesign or change in the side structure.

- Corporate cousin vehicles, if they are nearly identical except for their insignia and decoration (e.g., Ford Crown Victoria and Mercury Grand Marquis).

Merely sharing the same wheelbase or chassis does not make cars essentially identical by this definition – e.g., Buick LeSabre and Park Avenue, or the 1991 Pontiac Grand Am and the 1992 Pontiac Grand Am (same chassis, different body) are not essentially identical. The extended group of cars with known TTI(d) includes approximately 20-30 percent of the fleet in pre-standard model years 1981-93 (when only research tests were performed on a limited, but fairly representative cross-section of cars), but 80-90 percent of the fleet in post-standard vehicles (when compliance and NCAP tests were performed on a nearly comprehensive list of cars).

Figure 1-10 and the first column of Table 1-1 show the sales-weighted average TTI(d) of new passenger cars by model year, from 1981 to 2002.[33] They show a steady, almost linear record of improvement in TTI(d) from 95 to 60, with perhaps a slightly larger-than-usual drop in the early 1990's. (The second column of Table 1-1 shows the percent of that model year's sales for which TTI(d) is known.)

---

[33] Every "average TTI(d)" number in Tables 1-1 – 1-4 and Figures 1-10 – 1-12 is derived by taking a sales-weighted average of the make-models for which TTI(d) has been measured in a test (of that specific vehicle, or a nearby model year/corporate cousin of the same design). These numbers are subject to error, especially pre-FMVSS 214, because they may comprise relatively few distinct make-models and a relatively small percentage of all make-models sold that year.

FIGURE 1-10: AVERAGE TTI(d) BY MODEL YEAR, ALL PASSENGER CARS

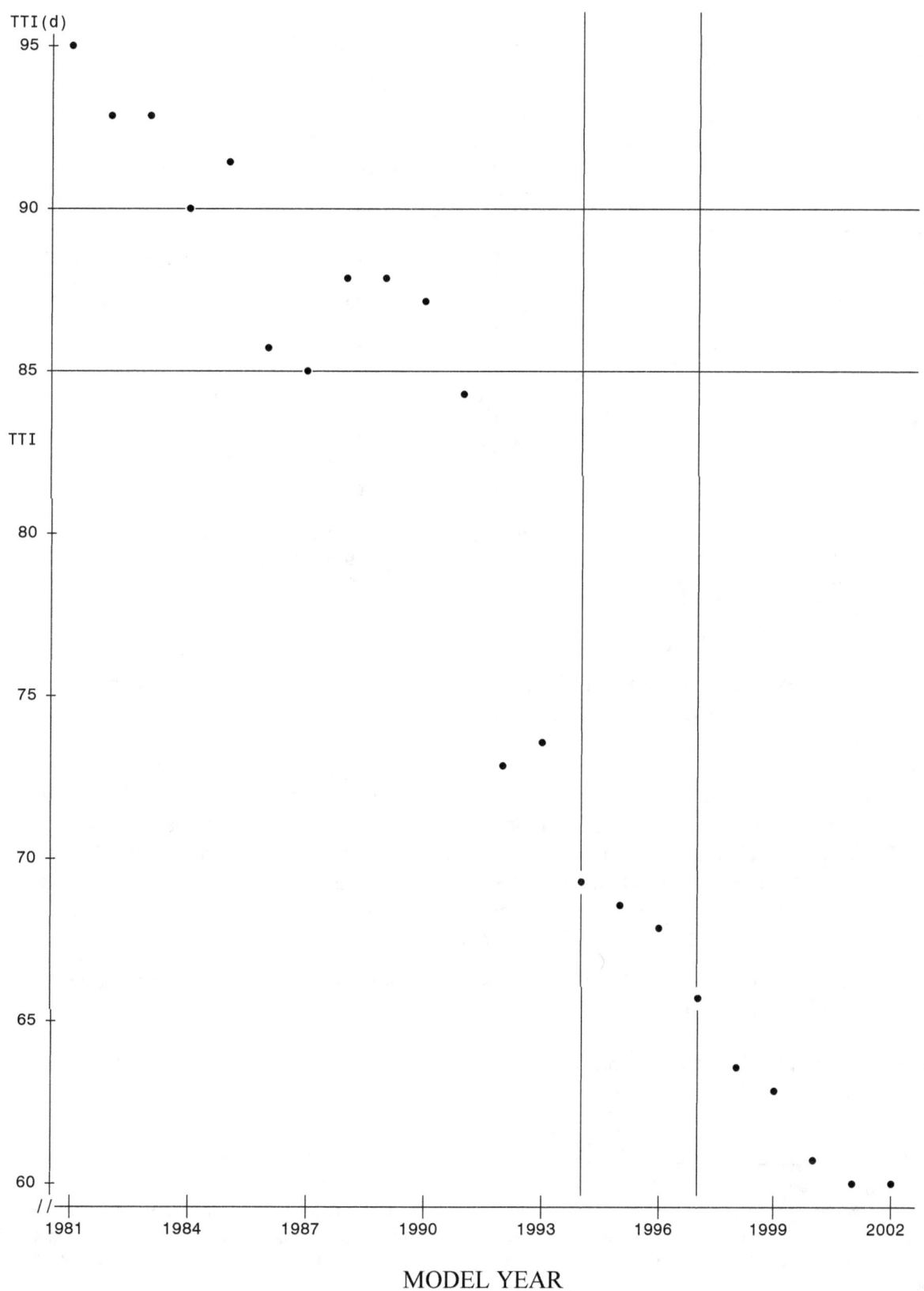

MODEL YEAR

21

TABLE 1-1: AVERAGE TTI(d) BY MODEL YEAR, PASSENGER CARS

MODEL YEAR	TTI ALL CARS	PCT OF CARS W KNOWN TTI	TTI 2-DOOR CARS	TTI 4-DOOR CARS
1981	95.1	17.8	108.3	87.9
1982	93.0	23.0	113.6	86.0
1983	92.6	24.6	117.9	85.5
1984	90.1	27.0	116.6	84.2
1985	91.2	30.1	113.5	83.7
1986	85.8	34.4	111.9	78.2
1987	85.3	34.4	106.6	79.0
1988	87.8	37.4	109.3	77.6
1989	87.7	35.4	109.5	76.3
1990	87.1	29.6	109.3	79.0
1991	84.4	23.4	112.1	78.5
1992	72.7	32.9	109.3	69.5
1993	73.6	39.3	91.4	69.9
1994	69.0	53.5	76.3	67.7
1995	68.4	69.3	76.8	65.8
1996	67.8	80.3	78.5	64.6
1997	65.8	91.8	72.1	64.1
1998	63.5	90.6	71.0	61.8
1999	63.1	89.0	70.1	61.5
2000	60.6	90.8	66.0	59.4
2001	59.9	83.5	63.2	59.2
2002	59.7	80.6	62.8	59.0

The overall results (in the first column of Table 1-1) are not that meaningful because they combine the effect of two separate trends:

- The market shift from 2-door to 4-door cars. In model year 1981, 2-door cars accounted for 50 percent of sales, but by 1998 they were less than 20 percent of the market.[34] Two-door cars were intrinsically more vulnerable than 4-door cars in side impacts, and historically had higher TTI(d), because the door of a 2-door car is usually much longer than the front door of a 4-door car. Impacting vehicles are less likely to strike pillar(s), more likely to hit the long, weakly supported door area between pillars.[35] The market shift to 4-door cars lowered overall, average TTI(d).

- In addition to the market shift between 2- and 4-door cars, TTI(d) also improved within 2-door cars and within 4-door cars.

Figure 1-10a and the last two columns of Table 1-1 indicate the trends in average TTI(d) separately for 2-door and 4-door cars.

---

[34] Tarbet, p. 121.

[35] John Jacobus suggested in his peer review of this report that the historically long, vulnerable doors of 2-door cars may have been shortened over the years during integrated platform redesigns or eliminated when some of the larger cars were discontinued as 2-door models. This may have contributed to the TTI(d) improvement in 2-door cars and helped bring their performance closer to 4-door cars.

FIGURE 1-10a: 2-DOOR VS. 4-DOOR CARS, AVERAGE TTI(d) BY MODEL YEAR

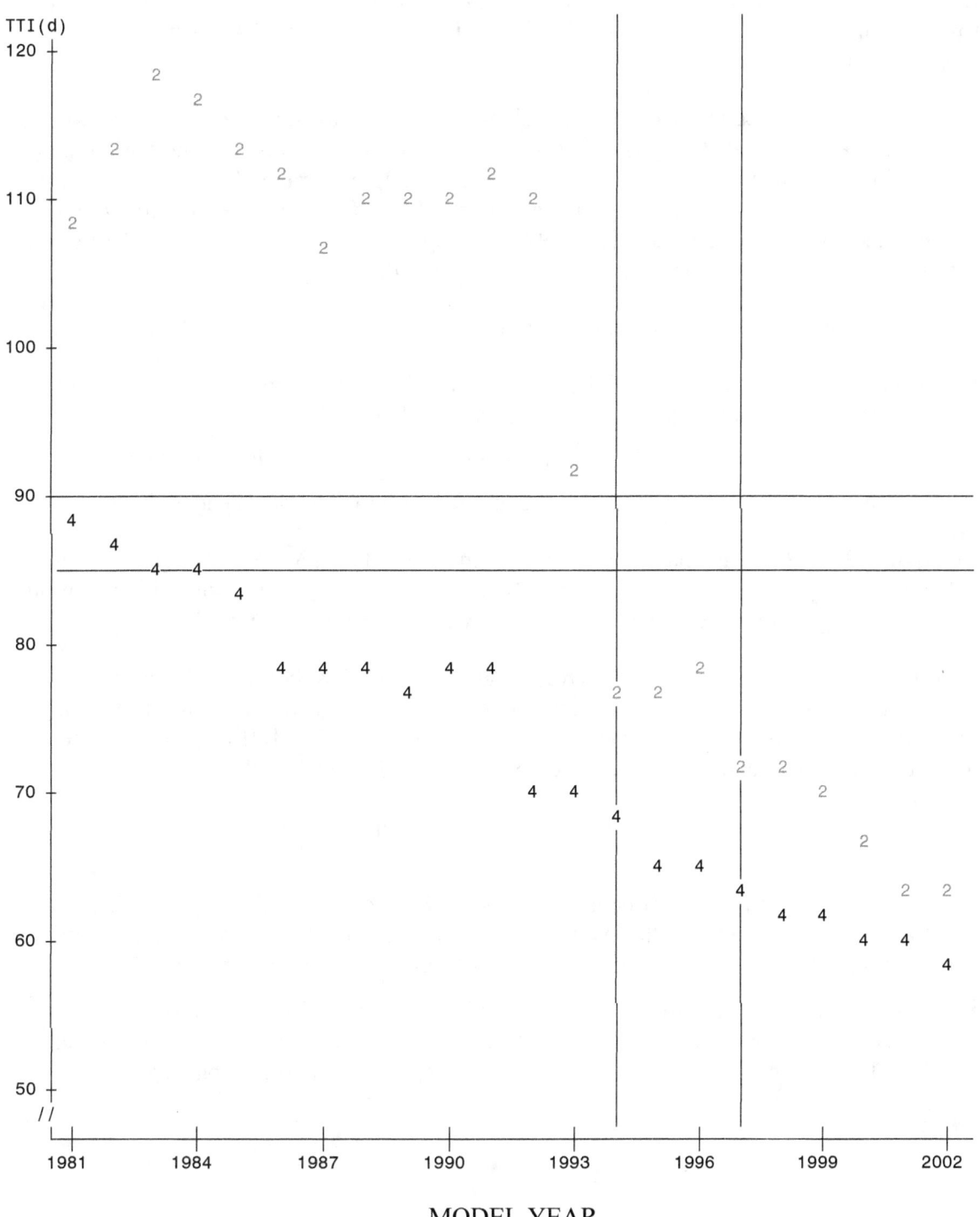

MODEL YEAR

TTI(d) in pre-standard 2-door cars originally averaged more than 110, well above the 90 that FMVSS 214 would subsequently allow. It appears that TTI(d) began to improve before the phase-in (although the exact timing here is uncertain, since only a limited number of cars were tested), and it dropped into the 70's during the phase-in period. Average TTI(d) continued to improve from the low 70's to the low 60's in 1997-2002, after FMVSS 214 was phased in, and nearly caught up to the performance level of 4-door cars.

By contrast, even at the beginning of the rulemaking process, the average 4-door car had TTI(d) close to 85, the level permitted by FMVSS 214. TTI(d) improved both before and during the phase-in period, but much less than in 2-door cars. For example, from 1990 to 1997, TTI(d) fell by 37 units in 2-door cars (from 109 to 72), but only by 15 units in 4-door cars (from 79 to 64). TTI(d) continued to improve slightly after 1997, reaching 59 in 2002; in all post-standard years, average TTI(d) was far below the 85 allowed by the standard.

More insight is obtained if the 2-door and 4-door cars are each subdivided into three subgroups:

1. Cars not certified to meet FMVSS 214, including all cars up to model year 1993, and 1994-1996 make-models not yet self-certified for FMVSS 214.
2. Cars certified to meet FMVSS 214[36], but not equipped with side-impact air bags.
3. Cars certified to meet FMVSS 214 and equipped with torso and/or head air bags.

Note that, in 1994-1996, some make-models had been certified for FMVSS 214 (group 1) and others not yet (group 0). More recently, some cars have been equipped with side air bags (group 2) while others have not (group 1), sometimes even within the same make-model.

Figure 1-11 and Table 1-2 compare the TTI(d) trends for the three subgroups of 2-door cars. TTI(d) remained fairly constant, over time, **within** each subgroup while differing substantially **between** subgroups. Thus, the overall trend of steadily declining TTI(d) (Figure 1-11) largely reflects the shift of more and more cars from group 0 to 1 to 2. Specifically:

- Pre-standard 2-door cars improved from the 110's in the early 1980's to the 90's in the mid-1990's.

- 214-certified cars without air bags were close to 70 throughout 1994-2002, with perhaps a drop into the mid-60's in the last two years. In other words, subgroup 1 is substantially lower than the later pre-standard cars, but has, itself, changed little over time.

- Cars with side air bags had TTI(d) close to 45 throughout 1998-2002. That is a substantial improvement of 25 units relative to 214-certified cars without side air bags, about the same magnitude as the improvement from subgroup 0 to subgroup 1.

---

[36] I.e., cars produced during the phase-in period (9/1/1993 – 8/31/1996) and self-certified by their manufacturers, plus all cars produced on or after 9/1/1996.

FIGURE 1-11: 2-DOOR CARS, AVERAGE TTI(d) BY MODEL YEAR

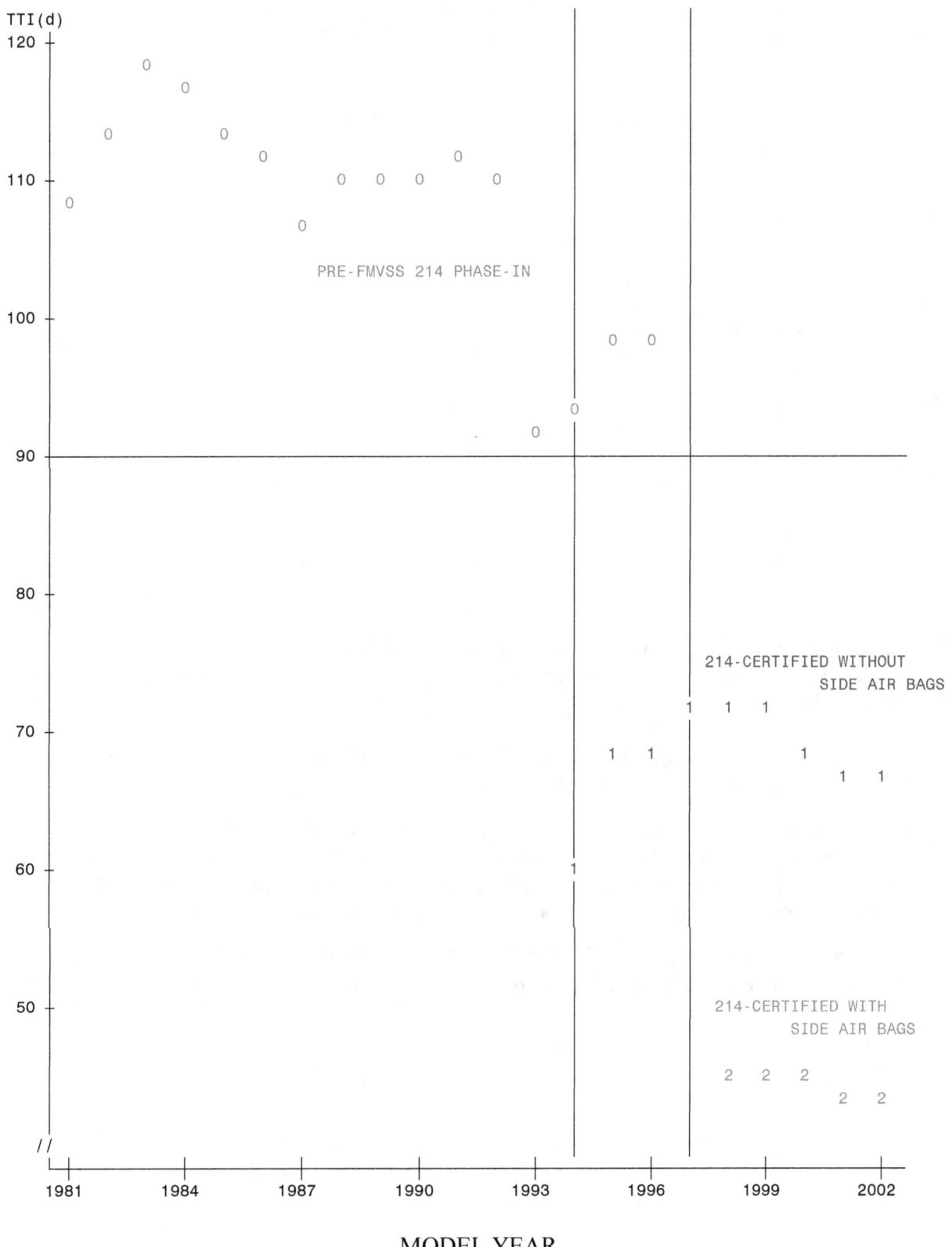

MODEL YEAR

TABLE 1-2: AVERAGE TTI(d) BY MODEL YEAR, 2-DOOR CARS

MODEL YEAR	ALL 2-DOOR	PRE-214	214 CERT W/O BAGS	214 CERT WITH SIDE AIR BAGS
1981	108.3	108.3	.	.
1982	113.6	113.6	.	.
1983	117.9	117.9	.	.
1984	116.6	116.6	.	.
1985	113.5	113.5	.	.
1986	111.9	111.9	.	.
1987	106.6	106.6	.	.
1988	109.3	109.3	.	.
1989	109.5	109.5	.	.
1990	109.3	109.3	.	.
1991	112.1	112.1	.	.
1992	109.3	109.3	.	.
1993	91.4	91.4	.	.
1994	76.3	93.0	60.7	.
1995	76.8	97.6	67.8	.
1996	78.5	98.1	68.8	.
1997	72.1	.	72.1	.
1998	71.0	.	71.8	44.4
1999	70.1	.	71.4	44.4
2000	66.0	.	68.5	44.3
2001	63.2	.	65.9	43.0
2002	62.8	.	66.6	44.1

Figure 1-12 and Table 1-3 illustrate the corresponding trends in 4-door cars. Here, too, TTI(d) remained almost constant over time in subgroups 1 and 2, while declining gradually in subgroup 0. Specifically:

- Pre-standard/non-certified 4-door cars gradually improved from about 85 in the early 1980's to the low 70's in the mid-1990's, a moderate improvement.

- 214-certified cars without air bags were close to 63 throughout 1994-2002. This is somewhat lower than the last generation of pre-standard cars. Unlike 2-door cars, there is no dramatic improvement, at any specific time, in the **overall average** TTI(d) of 4-door cars without air bags, although there is a sizable cumulative effect over time.

- Cars with side air bags had TTI(d) close to 48 throughout 1996-2002. That is a fairly substantial decrease of 15 units relative to 214-certified cars without side air bags.

FIGURE 1-12: 4-DOOR CARS, AVERAGE TTI(d) BY MODEL YEAR

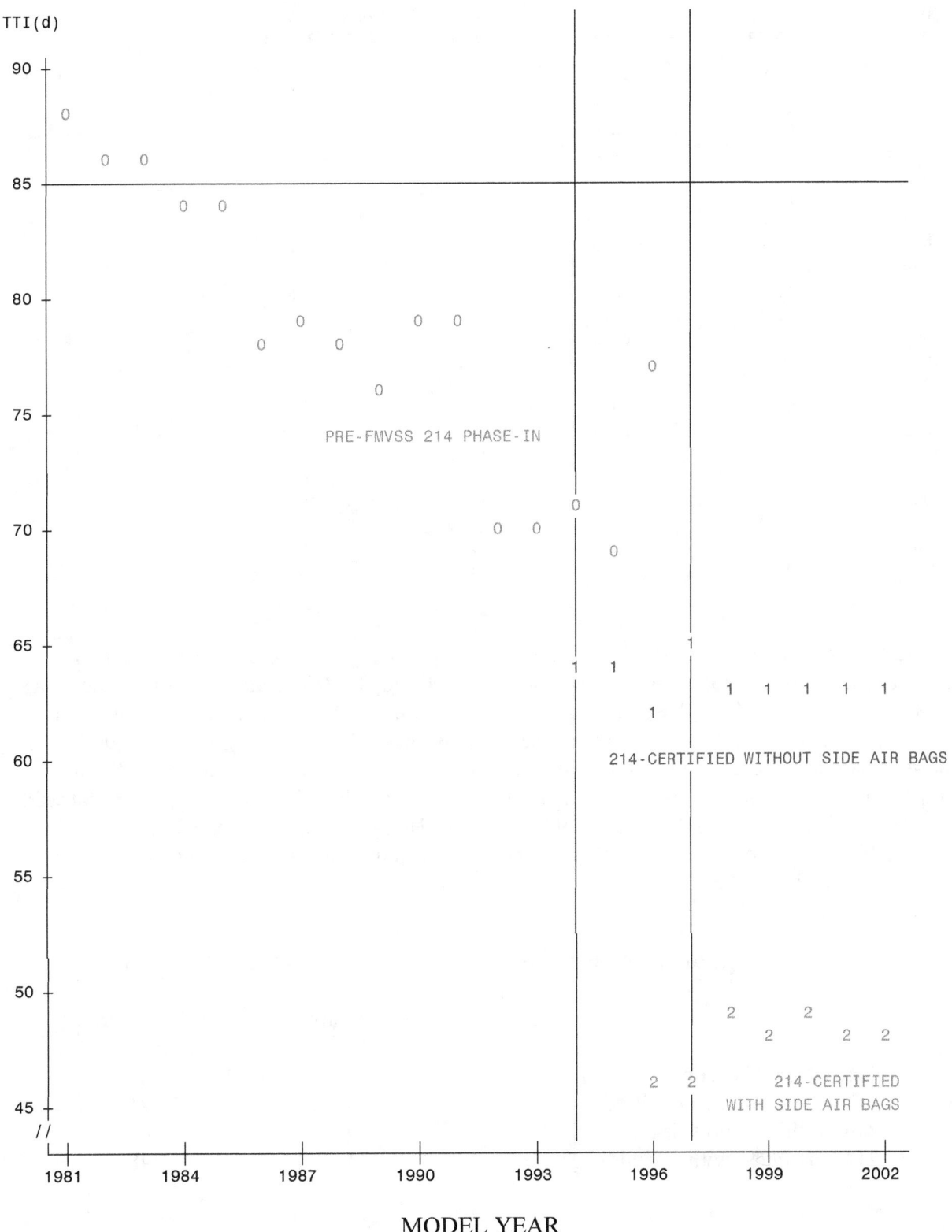

TABLE 1-3: AVERAGE TTI(d) BY MODEL YEAR, 4-DOOR CARS

MODEL YEAR	ALL 4-DOOR	PRE-214	214 CERT W/O BAGS	214 CERT WITH SIDE AIR BAGS
1981	87.9	87.9	.	.
1982	86.0	86.0	.	.
1983	85.5	85.5	.	.
1984	84.2	84.2	.	.
1985	83.7	83.7	.	.
1986	78.2	78.2	.	.
1987	79.0	79.0	.	.
1988	77.6	77.6	.	.
1989	76.3	76.3	.	.
1990	79.0	79.0	.	.
1991	78.5	78.5	.	.
1992	69.5	69.5	.	.
1993	69.9	69.9	.	.
1994	67.7	71.2	63.8	.
1995	65.8	69.2	63.6	.
1996	64.6	76.6	62.1	46.4
1997	64.1	.	64.5	45.8
1998	61.8	.	63.1	49.1
1999	61.5	.	63.2	48.3
2000	59.4	.	62.8	48.7
2001	59.2	.	62.9	48.1
2002	59.0	.	62.9	48.5

Table 1-4 summarizes these data by computing the sales-weighted average TTI(d) for four subgroups. Among 2-door cars that were not 214-certified, TTI(d) averaged 114 in model years 1981-1985 and 95 in 1993-1996, an improvement of 19 units. Presumably, the improvement was gradual from 1985 to 1993, but the exact trend is uncertain, since only a limited number of pre-standard cars were tested Upon 214 certification, average TTI(d) dropped to 69 (26 unit improvement), and upon installation of side air bags, to 44. The TTI(d) of 4-door cars that were not 214-certified averaged 85 in 1981-1985 and 71 in 1993-1996, an improvement of 14 units. Upon 214 certification, average TTI(d) dropped to 63 (8 unit improvement), and upon installation of side air bags, to 48.

TABLE 1-4: AVERAGE TTI(d) FOR SELECTED GROUPS OF CARS

	2-Door Cars	4-Door Cars
1981-1985 ("baseline TTI(d)")	114	85
1993-1996, not 214 certified	95	71
214-certified – no side air bags	69	63
214-certified – with side air bags	44	48

It is also interesting to compare the percentiles of distributions as well as the means:

	10th Percentile	Mean	90th Percentile
2-Door Cars			
Pre-standard/not 214-certified	86	108	119
214-certified without side air bags	58	69	81
214-certified with side air bags	37	44	48
4-Door Cars			
Pre-standard/not 214-certified	57	78	98
214-certified without side air bags	51	63	74
214-certified with side air bags	43	48	54

For 2-door cars, there are clear distinctions between all three subgroups. Only the best performers among pre-standard cars could have met FMVSS 214 (the 10th percentile of TTI(d) was 86). Poor performers among 214-certified cars (90th percentile was 81) did better than good pre-standard performers (10th percentile was 86). Likewise, even relatively poor performers with side air bags (90th percentile was 48) did better than excellent performers among 214-certified cars without the bags (10th percentile was 58).

By contrast, among 4-door cars, the distributions before and after 214-certification extensively overlap. The majority of pre-standard cars did better than the 85 eventually permitted by the future standard; in fact, the 75th percentile (not shown in the preceding table) was 81. Certification eliminated the worst performers (improving the 90th percentile score from 98 to 74) but had less effect on the good performers (improving the 10th percentile score from 57 to 51). Side air bags improved scores substantially, but even here there is some overlap between the least impressive cars with side air bags (90th percentile = 54) and the best cars without them (10th percentile = 51).

Fundamentally, some 4-door make-models have had good TTI(d) performance for a long time and did not change much in their design or in their test scores. These barely-changed models could obscure the attempt to identify the effect of 214-certification on fatality risk. Thus, the evaluation cannot simply compare the fatality rates of all 214-certified vs. all pre-standard cars, but will need to single out those make-models that experienced a substantial improvement in TTI(d). The addition of side air bags improved TTI(d) performance as much as, and sometimes more than the original certification to FMVSS 214; side air bags should be evaluated separately from the basic analysis of 214-certification.

1.6   Summary of the Phase 1 evaluation report

NHTSA completed Phase 1 of the evaluation of FMVSS 214 and published a report in 1999. The evaluation was limited to pre-standard cars (model years 1980-1992). It compared the side-impact fatality risk of cars with good TTI(d) scores (for that era) to cars with poor scores. TTI(d) scores were available for 43 make-model-model year combinations. The evaluation gathered 1980-1998 FARS cases of front-outboard occupant fatalities in each of these combinations, including cars from earlier/later model years and corporate "twins" of essentially identical design. "Side-impact fatality risk" is the ratio of occupant fatalities in side impact crashes (where the principal impact is at a 2-4:00 or 8-10:00 location) to fatalities in purely frontal crashes (where the principal impact is at a 12:00 location and the most harmful event is not a subsequent rollover).

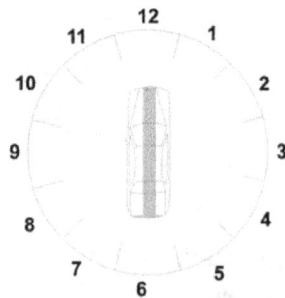

Purely frontal crashes were an acceptable control group because cars of the 1980's did not substantially change their technologies for occupant protection in frontal crashes (cars with frontal air bags or automatic belts were excluded from the analysis). The data file comprised 10,983 cases of occupant fatalities in side impacts and 12,019 in frontals.[37] Separate analyses were performed for the 2-door cars and the 4-door cars. Analysis techniques included:

- Correlation of TTI(d) with side-impact fatality risk across the various make-model-MY groups.[38]

- Comparing the fatality risk of the cars with the best TTI(d) scores to the cars with the poorest scores.[39]

- Logistic regression of the probability that a fatality is in a side impact (and not in a frontal), as a function of TTI(d), curb weight, driver age and other variables.[40]

In the 2-door cars, all three analysis methods showed significantly lower fatality risk in the cars with lower TTI(d) scores. For example, cars with TTI(d) > 115 had more fatalities in side impacts than in pure frontals, whereas cars with TTI(d) < 102 had substantially fewer[41]:

---

[37] Kahane (1999), pp. 11-18 and 24-31.
[38] *Ibid.*, pp. 31-45.
[39] *Ibid.*, pp. 47-71.
[40] *Ibid.*, pp. 73-103.
[41] *Ibid.*, p. 51.

2-DOOR CARS	Purely Frontal Fatalities	Side Impact Fatalities
Models with TTI > 115 (average = 123)	760	812
Models with TTI < 102 (average = 94)	654	528

This is a statistically significant 24 percent reduction of side impact fatalities with lower TTI(d): $1 - [(528/654) / (812/760)] = .244$. More precisely, a logistic regression estimated that fatality risk was reduced by a statistically significant 0.927 percent per unit decrease in TTI(d). That effect was calibrated from crash data for 17 2-door make-models of model years 1981-1993; none were FMVSS 214-certified; 16 of the 17 make-models had TTI(d) over 90. Importantly, unlike the preceding results on side door beams (Section 1.2), this reduction was not limited to single-vehicle crashes, but was about equally large in multivehicle and single-vehicle crashes. In fact, the highest effect was observed for nearside occupants of 2-door cars that were hit in the side by another passenger car.[42]

Corresponding analyses of 4-door cars, however, did not show statistically significant relationships between TTI(d) and side-impact fatality risk. Unlike the 2-door cars, a large portion of the crash data involved cars with TTI(d) scores clustered in a relatively narrow range, making it statistically more difficult to find correlations between TTI(d) and risk. Direct comparisons of the fatality risk in 4-door cars with the best and worst TTI(d) scores (analogous to the preceding table for 2-door cars) showed 5-15 percent fatality reductions for the cars with the better scores, but these reductions were not statistically significant.[43]

1.7 Braver-Kyrychenko and McCartt-Kyrychenko analyses of side air bags

Braver and Kyrychenko published the first statistical analysis of crash data on side air bags in 2003, based on calendar year 1999-2001 data from FARS and the General Estimates System (GES) of NASS.[44] They compared driver fatality rates per 1,000 nearside crash involvements – with torso bags plus head protection (torso bags with separate curtains or tubes for head protection, or torso/head combination bags), with torso bags only, and without side air bags – in model year 1997-2002 passenger cars. These early data showed a statistically significant 45 percent fatality reduction for torso bags plus head protection and an 11 percent reduction for torso bags alone.

In 2006, McCartt and Kyrychenko revised the analysis and substantially updated it with FARS and GES data through CY 2004.[45] Passenger car analyses comprise MY 1997-2002 cars in CY 1999-2001 and MY 2001-2004 in CY 2000-2004; SUVs are analyzed for MY 2001-2004 in CY

---

[42] *Ibid.*, pp. viii, 84 and 91-100; $1 - (1 - .00927)^{110-82} = 23$ percent.

[43] *Ibid.*, pp. ix, 37-45, 63-71 and 85-87.

[44] Braver, E.R. and Kyrychenko, S.Y., *Efficacy of Side Airbags in Reducing Driver Deaths in Driver-Side Collisions*, Insurance Institute for Highway Safety, Arlington, VA, 2003; *Status Report*, Vol. 38, August 26, 2003, Insurance Institute for Highway Safety, Arlington, VA.

[45] McCartt, A.T. and Kyrychenko, S.Y., *Efficacy of Side Airbags in Reducing Driver Deaths in Driver-Side Car and SUV Collisions*, Insurance Institute for Highway Safety, Arlington, VA, 2006.

2000-2004. Driver fatality rates per 1,000 nearside crash involvements with torso bags plus head protection or with torso bags only are compared to the rates without side air bags. The authors adjusted the results by also comparing the corresponding fatality rates in a control group of frontal and rear-impact crashes. Side air bags are effective if they reduce fatality risk to a larger extent in nearside impacts than they "reduce" it in frontal and rear impacts. In other words, the data and analysis are quite similar although not identical to Chapter 3 of this report (which also relies on FARS and GES data and also compares risk nearside vs. frontal or rear impacts).

Torso bags plus head protection reduced drivers' fatality risk in nearside impacts by 37 percent in passenger cars and 52 percent in SUVs, relative to drivers without side air bags. Torso bags alone reduced fatality risk by 26 percent in cars and 30 percent in SUVs. All of the reductions are statistically significant.

McCartt's 37 and 26 percent reductions in passenger cars are higher than our best estimates in Section 3.10 of this report: 24 and 12 percent. Nevertheless, the 37 percent **exactly** matches our result, in Table 3-6a, on the analysis that most closely resembles McCartt's. However, Chapter 3 of our report also analyzes basically the same data by other techniques that consistently produce lower estimates of fatality reduction (averaged for our "best" estimate). Specifically, analyses in Chapter 3 that compare fatality risk with torso bags and head protection only to the risk in cars **of the same make-models** without side air bags produce lower estimates. So do analyses in Chapter 3 based on FARS data alone. In short, the McCartt-Kyrychenko study and this report, analyzing much of the same data by similar methods, strongly agree that torso bags plus head protection significantly reduces fatality risk in nearside impacts, but McCartt's estimate is near the top of the range of point estimates presented in Chapter 3 of this report.

1.8   Evaluation goals

The Government Performance and Results Act of 1993 and Executive Order 12866 (October 1993) require agencies to evaluate their existing programs and regulations. Evaluations determine the actual benefits – e.g., lives saved – and costs of safety equipment installed in production vehicles in connection with a rule. NHTSA's evaluation plan for 2004-2007 schedules and summarizes the agency's proposed evaluations, including this one.[46]

NHTSA has evaluated of the cost of FMVSS 214 in passenger cars. Modifications to meet the static strength requirement of 1973 and the dynamic impact standard in 1994-1997 added an average of $154 (in 2002 dollars) and 54 pounds of weight to 2-door cars. They added an estimated $187 and 64 pounds to 4-door cars. These averages take into account that nearly half the cars received only padding and/or minor structural changes, or even no changes at all to meet the dynamic standard. Studies of the cost of side-impact air bags are underway.[47]

The Phase 1 evaluation report showed a statistically significant association, in 2-door cars of the 1980's, between TTI(d) and fatality risk in actual side impacts, amounting to a 0.927 percent fatality reduction per unit decrease in TTI(d). This relationship, however, should not be

---

[46] Government Performance and Results Act of 1993, Public Law 103-62, August 3, 1993; *Federal Register* 58 (October 4, 1993): 51735; *NHTSA Evaluation Program Plan, 2004-2007*, p. 5.
[47] Tarbet, pp. 113-126.

generalized to estimate the effect of FMVSS 214 except, at most, as a first approximation. The fleet of new passenger cars during and after the 1994-1997 implementation of FMVSS 214 differs from the Phase 1 data in several important respects:

- TTI(d) had dropped from the 90-130 range to the 50-90 range. The fatality reduction per unit decrease in TTI(d) in this lower range, if there is any, could well be different from 0.927 percent.

- 80 percent of late-model passenger cars are 4-door.

- The vehicle modifications in the 1990's are different from the 1980's. For example, they include padding and, later, side air bags.

Essentially, we need to make a fresh start. The goals of this evaluation report are:

1. Track the TTI(d) history of **specific** make-models of passenger cars during and after the phase-in of FMVSS 214 (as opposed to Section 1.5, which tracked the sales-weighted average of all cars). Identify models that substantially (e.g., by 10 or more units) improved their TTI(d), **without side air bags**, when they were initially certified to FMVSS 214, or in some subsequent model year. Describe the modifications in those models. Also identify models whose TTI(d) essentially did not change upon 214-certification, and describe their modifications, if any.

2. Using crash data, statistically analyze the change in nearside-impact fatality risk for the models that substantially improved TTI(d). We will analyze nearside fatalities per 1,000 crash-involved occupants or the ratio of nearside fatalities to non-occupant fatalities (pedestrians and bicyclists), possibly controlling for factors such as vehicle age. As a check, also analyze the change in fatality risk among the models that 214-certified with little or no change in TTI(d).

3. Identify make-models of cars that began to offer torso air bags some years after they initially certified to FMVSS 214. Statistically analyze the difference in fatality risk, for these specific models, in the 214-certified cars with and without the torso air bags. Estimate the fatality reduction for torso air bags above and beyond the effects of the initial 214-certification.

4. Identify car models that offer head-protection air bags. Estimate the fatality reduction for torso plus head air bags above and beyond the effects of initial 214-certification. Also analyze the effect of head air bags (and torso air bags) on occupant ejection in crashes.

5. Estimate the combined effect of all measures since 1985 to reduce fatality risk in side impacts of passenger cars: the structural modifications/padding introduced before, during or shortly after the phase-in of FMVSS 214, torso air bags, and head-protection air bags.

The evaluation focuses more on passenger cars than on LTVs, because substantially more data are available for cars. But FMVSS 214 also applies to LTVs up to 6,000 pounds Gross Vehicle Weight Rating (GVWR), including, since September 1, 1998, a dynamic test requirement in which TTI(d) may not exceed 85 (same as 4-door cars).[48] In model years 1999-2003, all full-size vans offered for sale in the United States had GVWR over 6,000 pounds. In other words, the regulation only affected pickup trucks, SUVs and minivans – and only those with GVWR up to 6,000 pounds. NHTSA has little evidence that full-sized pickup trucks, SUVs and minivans required or received substantial modifications in structure or padding to meet FMVSS 214.[49] The relatively high, rigid floors of these LTVs, plus the side door beams already in the vehicles (see Section 1.2) were adequate for compliance with the dynamic test. On the other hand, the side structures of compact pickup trucks may have been upgraded in response to the dynamic test requirement, or as part of an Integrated Platform Redesign a year or more before the requirement. Section 2.5 of this report analyzes the limited crash data on compact pickup trucks.[50]

Torso air bags began to appear on some SUVs and minivans in 1998 and head-protection air bags in 1999. Introduction of side air bags has been slower than in passenger cars; moreover, there are fewer crash data because LTVs are less vulnerable to side impact than cars (see Figure 1-1). Section 3.7 analyzes the limited data on side air bags in LTVs.

---

[48] *Federal Register* 60 (July 28, 1995): 38749.

[49] In response to Information Requests (IR), the agency received descriptions of side-impact protection on three MY 1999 SUVs and four minivans that were compliance-tested. Two explicitly state that the LTV was unchanged from previous model years; three do not indicate any changes; one shows minor changes; and one provides extensive description of side structures but do not explain if any of them are new or all are carried over from the previous model year.

[50] John Jacobus recommended analyses of compact pickup trucks in his review of this report, and also supplied information on side structures in compact pickup trucks.

CHAPTER 2

EFFECT OF TTI(d) IMPROVEMENTS ON NEARSIDE FATALITIES IN MULTIVEHICLE CRASHES – AFTER 1993, WITHOUT SIDE AIR BAGS

2.0 Summary

NHTSA test results identify 15 make-models of passenger cars that substantially improved their TTI(d) test performance, without the addition of side air bags, at some point in 1994-2002 (during or after the phase-in of FMVSS 214). TTI(d) improved by an average of 23 units. Statistical analyses show a significant 18 percent reduction in the fatality risk of nearside, front-seat occupants when other vehicles strike these cars in the side.

2.1 A file of side impact test results for passenger cars

The starting point for the analysis is to compile the measurements of TTI(d) for correctly belted Side Impact Dummies (SID) in the front seat in side impacts to production passenger cars in the FMVSS 214 test configuration: an impact by a Moving Deformable Barrier (MDB) at an angle of 63.43 degrees with respect to the longitudinal centerline of a stationary test vehicle. As explained in Section 1.3, the wheels of the MDB are "crabbed" 26.57 degrees toward the rear of the test vehicle to obtain a right-angle contact and simulate an MDB moving x mph hitting, at a right angle, the door area of a subject vehicle, traveling x/2 mph.[51] The MDB simulates a 3000-pound passenger car. These impacts have been conducted, over the years, at speeds ranging from somewhat under 33.5 mph up to 38.5 mph. To compare the results of different tests, it is necessary to adjust TTI(d) for the test speed. The adjusted TTI(d), the value that would likely have been observed if the test had been run at the 33.54 mph velocity specified in FMVSS 214, as discussed in Section 1.5, is:

$$\text{TTI(d) adjusted} = \text{TTI(d) observed} * (33.54/\text{SPEED})^2$$

From 1981 through the end of the model year 2002 test program, results are available to NHTSA for 386 individual cars, comprising 318 distinct make-model-year-body style combinations (i.e., for certain combinations, two or more vehicles were tested).[52] These 386 tests include:

- 73 tests of pre-FMVSS 214 cars comprising 60 distinct make-model-year-body style combinations, including:

---

[51] x is close to 30 mph in the FMVSS 214 compliance test and close to 35 mph in the NCAP test.

[52] Not included are tests to non-production cars (e.g., experimentally modified with padding or structures not available in production cars that model year), or not in the FMVSS 214 configuration (e.g., pole impacts) or where TTI(d) was not measured, or unsuccessfully measured for a SID in the front-seat.

- 43 "research" tests sponsored by NHTSA or Transport Canada, usually at speeds slightly above 33.54 mph[53], in three series:

  - 20 MY 1980-1985 cars in the initial research leading up to the FMVSS.

  - 12 tests of MY 1986-1990 cars to support the regulatory analysis at the time the FMVSS was proposed

  - 11 tests of MY 1992-1993 cars to get a final baseline just before the standard's phase-in period

- 23 tests of 1988-1990 cars performed by manufacturers and considered in NHTSA's regulatory analysis, among them:

  - 14 baseline tests of different MY 1988 vehicles, originally submitted to NHTSA as confidential information. Subsequently, two manufacturers permitted NHTSA to disclose the individual results.

  - A total of 9 tests on two 1990 make-models to investigate the repeatability of the test, submitted to NHTSA as public information.

- 7 tests of pre-Standard 1995-1996 cars, performed especially for this evaluation to learn the TTI(d) performance of seven high-sales make-models just before certification. Similar to compliance tests, speeds were slightly less than 33.54 mph. The results were:

	Test Speed	Observed TTI(d)	Adjusted TTI(d)
1995 4-door Ford Taurus	33.0	70.	72.3
1995 4-door Saturn SL2	33.1	67.	68.8
1995 4-door Ford Escort	33.2	66.	67.4
1996 4-door Pontiac Grand Am	33.0	62.	64.0
1995 2-door Chevrolet Cavalier	33.0	107.	110.5
1995 4-door Honda Civic	32.9	51.5	53.5
1995 4-door Buick LeSabre	32.6	68.2	72.2

- 313 tests of 1994-2003 cars certified to meet FMVSS 214 comprising 258 distinct make-model-year-body style combinations, including:

  - 122 NHTSA-sponsored compliance tests at speeds slightly less than 33.54 mph.

---

[53] The actual speed in MDB impacts is not necessarily the exact speed intended by the testing facility, but typically varies by some fraction of one mph due to tolerances in the machinery. Therefore, manufacturers (on self-certification tests) and researchers typically target a speed slightly higher than 33.54 mph to assure that actual speed, allowing for error, is no less than 33.54 mph – because a passing score above 33.54 mph may be presented as evidence that the vehicle would also pass at exactly 33.54 mph. Conversely, compliance tests are usually targeted slightly below 33.54 mph to assure that actual speed, allowing for error, does not exceed 33.54 mph – because a failing score below 33.54 mph may be presented as evidence that the vehicle would also fail at exactly 33.54 mph. However, the actual speed never differed from 33.54 by more than one mph and, in any case, the speed adjustment formula compensates for the differences in actual test speeds.

- 119 NHTSA-sponsored tests at approximately 38.5 mph to provide information for NCAP, and sometimes also serving as compliance tests.[54]

- 56 tests by General Motors on 1994-1997 cars that they subsequently certified to meet FMVSS 214. GM included the test results in their public comment on NHTSA's Phase 1 evaluation report (Docket No. NHTSA-99-6545-05) in response to the agency's request for test results to support this phase of the evaluation. Speeds were slightly higher than 33.54 mph.

- 16 tests by manufacturers of 1997-2002 cars at 38.5 mph or slightly higher, submitted to NHTSA as public information, to supplement or complement the NCAP results from the NHTSA-sponsored tests.

2.2 Make-models that substantially improved TTI(d) without side air bags

The next steps are to sort the test results by make-model, track each make-model's TTI(d) results and side-structure design over the years, and identify groups of make-models that substantially improved their TTI(d) (without side air bags) and/or received major structural upgrades at some point. These are the models where we have the best chance of observing a statistically significant reduction of side-impact fatalities after TTI(d) was improved. Likewise, we will identify other make-models whose TTI(d) has not changed over time. They may serve as a control in some of the analyses.

We will define 40 "make-model groups" that consist of a single make-model over several model years (during which time it may be redesigned), or of two or more make-models that, in any given model year, are quite similar vehicles (such as Ford Taurus and Mercury Sable). Two-door and 4-door cars of the same make-model must go into separate groups. The Phase 1 evaluation report for FMVSS 214, specifically its Chapter 8 and Appendix B already defined most of the groups. We have dropped 15 of the 52 groups defined in the Phase 1 report because little or no TTI(d) history is available or because the make-model was discontinued before FMVSS 214 certification, but have added three groups whose test histories indicate a substantial improvement in TTI(d), without air bags, at some point in time.[55]

Appendix A of the current report defines and discusses each of the 40 make-model groups. For example, one group includes Ford Taurus and the quite similar Mercury Sable. Here are their test results over time, as measured by speed-adjusted TTI(d). (When two or more cars were tested in the same year, the harmonic mean[56] of the adjusted TTI(d)'s is calculated.):

---

[54] If the TTI(d) and pelvic g's for both dummies in these 38.5 mph tests were below the levels specified in FMVSS 214, NHTSA accepts the results as evidence of compliance with FMVSS 214 and does not retest at 33.54 mph.

[55] Kahane, C.J., *Evaluation of FMVSS 214 - Side Impact Protection: Dynamic Performance Requirement; Phase 1: Correlation of TTI(d) with Fatality Risk in Actual Side Impact Collisions of Model Year 1981-1993 Passenger Cars*, NHTSA Technical Report No. DOT HS 809 004, Washington, 1999, pp. 139-155 and 185-238.

[56] $\exp((\log TTI_1 + \ldots + \log TTI_n) / n)$. For statistics such as test scores that, when they are high can be very high (distribution skewed to the right), the harmonic mean is a better indicator of the central tendency than the arithmetic mean, because it gives less weight to the anomalously high scores. When there is little difference between scores (excellent repeatability), the harmonic and arithmetic means are nearly identical.

Model Year	Adjusted TTI(d)	N of Tests	Side Air Bags?	Car Body Platform (Wheelbase and Other Shared Features)
1988	78.2	1	no	1986-95 FORD TAURUS 106 wheelbase
1990	77.4	7	no	1986-95 FORD TAURUS 106 wheelbase
1995	72.3	1	no	1986-95 FORD TAURUS 106 wheelbase
1996	50.7	1	no	1996- FORD TAURUS 108.5 wheelbase
1997	57.1	1	no	1996- FORD TAURUS 108.5 wheelbase
1999	61.7	1	no	1996- FORD TAURUS 108.5 wheelbase
2000	64.7	1	no	1996- FORD TAURUS 108.5 wheelbase

Taurus and Sable were substantially redesigned on a new body platform in 1996. At that time, Ford also initially certified that they met FMVSS 214. In fact, the TTI(d) history suggests that even the first-generation (1986-1995) Taurus had scores in the 70's, better than the 85 allowed by FMVSS 214 for 4-door cars. However, there is a clear improvement in the 1996 redesign, with TTI(d) dropping into the 50's in 1996-1997.[57] Furthermore, in response to NHTSA's Information Request (IR), Ford reported that the 1996 Taurus received structural reinforcements in the body side structure and energy-absorbing foam in the door panels. However, there is no accompanying picture showing the scope of these reinforcements, nor it is entirely clear what the word means, given that the entire side structure may have been altered as part of the 1996 major redesign of the Taurus. Still, the most likely explanation of what happened is:

- TTI(d) was close to 72 in the first-generation Taurus/Sable, up to 1995. When these cars were redesigned and certified to FMVSS 214 in 1996, they received major structural modifications and padding, and TTI(d) improved to about 54 in 1996-97.

The objective is to assign each of the 40 make-model groups to one of four larger categories based on the history of their TTI(d) test results and on their vehicle modifications, as documented by their manufacturers in response to the Information Requests (IR):

1. Make-model groups that substantially improved TTI(d) (namely, by 9 or more units) upon FMVSS 214 certification or at some other time, without side air bags. When we assign a make-model group to this category we must also specify when TTI(d) improved. The IR should be consulted for a description of how the vehicle changed. When IR's are available and complete for these make-models, they usually show major new structure. It is also useful to specify if the TTI(d) improvement coincided with an overall vehicle redesign; if not, we have the special case of a vehicle essentially unchanged except for the steps taken to improve TTI(d).

2. Make-model groups with no TTI(d) change or limited change, as evidenced by:

   o Actual TTI(d) history showing little or no change, or

   o IR unequivocally stating that the car certified to FMVSS 214 without any changes from the preceding model year.

---

[57] The 1999 and 2000 results were slightly higher; presumably this is merely test-to-test variation, as there is no evident reason for performance to have deteriorated.

3. Make-model groups where the IR perhaps suggests they received major new structure upon FMVSS 214 certification; however, NHTSA does not know the amount of TTI(d) change, if any, because the agency has no test data for a car of the model years immediately preceding certification – i.e., there is no demonstration of a TTI(d) improvement to corroborate the suggestion that structure was upgraded.

4. Other make-model groups with unknown TTI(d) change upon FMVSS 214 certification (because the agency has no pre-certification test data), such as:

   o Models whose IR explicitly states they received only padding and/or minor structural change, or

   o IR does not clearly specify what happened, or

   o There is no IR.

In our example, Taurus/Sable may be assigned to Category 1, substantial TTI(d) improvement without air bags upon original FMVSS 214 certification in 1996, as evidenced by a decrease in TTI(d) from about 72 to about 54 (18 units). The improvement coincided with a major redesign, and it apparently involved major structural reinforcement plus the addition of padding.

Table 2-1a, based on Appendix A, lists the 15 make-model groups in Category 1, indicating the model year in which NHTSA believes TTI(d) substantially improved; the "before" and "after" TTI(d) and the difference between them; and the vehicle modifications in that year (if known). Table 2-1a also indicates if the TTI(d) improvement coincided with an overall redesign of that make-model, and if it occurred in the same year as the original FMVSS 214 certification.

The decrease in TTI(d) ranged from 9 units in the 1998 Honda Accord 2-door up to 49 units in the 1997 Chevrolet Corvette. The registration-weighted[58] average improvement was 23 units of TTI(d). Eight of the 15 groups substantially improved TTI(d) when they were originally certified to meet FMVSS 214; six, in a later model year; Ford Mustang, apparently two years before certification.

---

[58] In Tables 2-1a and 2-1b, each of the 15 make-model groups is weighted by the total vehicle registration years in calendar years 1993-2004 for cars in the model year ranges shown in Table 2-1b ("before" plus "after"). The weights correspond to their relative contributions to the crash data analyzed later in this chapter.

TABLE 2-1a: CATEGORY 1 MAKE-MODEL GROUPS THAT SUBSTANTIALLY IMPROVED TTI(d) UPON FMVSS 214 CERTIFICATION OR AT SOME OTHER TIME, WITHOUT SIDE AIR BAGS

Make-Model(s)	MY of TTI(d) Improvement	TTI(d) Before	TTI(d) After	Δ	Vehicle Changes	Integrated Platform Redesign?	MY of 214 Cert.
Dodge Intrepid/Concorde/Vision (4 door)	1994	79	65	14	Maj.Str.	No	Same
Ford Mustang (2 door)	1994	110	63	47	Unknown	Yes	1996
Ford Taurus/Mercury Sable (4 door)	1996	72	54	18	Maj.Str.+Pad	Yes	Same
Chevrolet Corvette (2 door)	1997	109	60	49	Unknown	Yes	Same
Chevrolet Cavalier/Pontiac Sunfire 2 door	1997	111	81	30	Pad+Min.Str.	No	Same
Chevrolet Monte Carlo (2 door)	2000	77	63	14	Unknown	Yes	1995
Pontiac Grand Am/Achieva/Skylark 2 door	1997	109	70	39	Unknown	No	Same
Nissan Sentra 4 door	1995	92	66	26	Maj.Str.+Pad	Yes	Same
Honda Civic 2 door	1996	86	71	15	Maj.Str.	Yes	Same
Honda Accord 2 door	1998	72	63	9	Unknown	Yes	1994
Honda Accord 4 door	1998	77	59	18	Maj.Str.	Yes	1994
Subaru Legacy 4 door	2000	68	48	20	Unknown	Yes	1995
Subaru Impreza 4 door	2002	72	45	27	Unknown	Yes	1994
Toyota Corolla/Geo Prizm (4 door)	1997	91	66	25	Maj.Str.+Pad	No	Same
Mitsubishi Eclipse 2 door	2000	89	52	37	Unknown	Yes	5/1995

REGISTRATION-WEIGHTED AVERAGE TTI(d) IMPROVEMENT: 23

40

It can be unclear what types of vehicle modifications brought about the TTI(d) improvements observed in the tests. There may be no IR at all (especially if the improvement was not in the year of FMVSS 214 certification) or it may be difficult to interpret the IR.[59] For the seven groups with relatively unambiguous IR's, six received major new or reinforced structures, and only one (Chevrolet Cavalier 2-door) was primarily limited to padding.

The analysis has to concentrate on these 15 groups – the only make-models with hard evidence that TTI(d) decreased – and that has several disadvantages. Ideally, we would have liked make-models where essentially nothing else changed in the year that side impact protection was added and TTI(d) decreased. Table 2-1a, however, shows that TTI(d) improvement was part of an integrated platform redesign of the car in 11 of the 15 groups. When cars are redesigned, fatality rates may change, for example, because the redesigned vehicle attracts a different type of driver.

TTI(d) values are based on single or occasionally multiple tests and, as such, are a statistic subject to measurement error. Results can vary from test to test especially if the vehicles are not exactly identical (e.g., two different model years of basically the same car), or the speeds are not the same (the speed adjustment factor works on the average, but certain individual make-models can have lower or higher NCAP results than might be expected, based on their compliance test results). An exceptionally high reading in one model year followed by an exceptionally low one in a later year could create the appearance of a substantial TTI(d) improvement that did not really happen. Specifically, there could be some questions about the results for the Dodge Intrepid[60] and the Chevrolet Cavalier.[61]

Finally, when there are just 15 make-models, and much of the exposure is concentrated in the high-sales models at the top of the list, the data are fairly "clustered" and could have, in a sense, less statistical power than what would be suggested by the N of crash cases. If by coincidence Ford Taurus, Honda Accord and Toyota Corolla, each for their own particular reasons unrelated to the side structure, all happened to experience a fatality reduction in the years TTI(d) decreased, we are almost guaranteed a positive result for the analysis because these three models constitute about half of the data. Analysis results will need to be viewed with a little more than the usual skepticism, and corroborated by alternate methods involving a variety of controls.

Table 2-1b shows the model years for the 15 groups that will be included in the analyses.

---

[59] For reasons such as: (1) no diagrams; (2) IR not for the same model year as the TTI(d) decrease; (3) if vehicle was redesigned, not clear if IR covers improvements "built into" the new design; (4) ambiguity if structures discussed in the IR actually changed from the previous year.

[60] The IR describes extensive structures in the 1994 Dodge Intrepid and that's consistent with a TTI(d) of 79 in 1993 and 65 in 1994. Nevertheless, it is surprising that an entirely new car would be introduced in 1993 and then extensively modified the next year. Perhaps these structures had already been built into the original 1993 Intrepid, and the relatively high test score for the 1993 specimen was merely an anomaly.

[61] Although the TTI(d) of 111 in 1995 and 81 in 1997 are well documented, it is surprising that such a large improvement was achieved with just padding and some minor structural enhancements.

TABLE 2-1b: CATEGORY 1 MAKE-MODEL GROUPS THAT SUBSTANTIALLY IMPROVED TTI(d) MODEL YEARS INCLUDED IN THE EVALUATION

| Make-Model(s) | "Before" | | "After" | | Δ TTI(d) |
	Model Years	TTI(d)	Model Years	TTI(d)	
Dodge Intrepid/Concorde/Vision (4 door)	1993	79	1994-1996	65	14
Ford Mustang (2 door)	1991-1993	110	1994-1996	63	47
Ford Taurus/Mercury Sable (4 door)	1993-1995	72	1996-1998	54	18
Chevrolet Corvette (2 door)	1994-1996	109	1997-1999	60	49
Chevrolet Cavalier/Pontiac Sunfire 2 door	1995-1996	111	1997-1999	81	30
Chevrolet Monte Carlo (2 door)	1997-1999	77	2000-2002	63	14
Pontiac Grand Am/Achieva/Skylark 2 door	1994-1996	109	1997-1998	70	39
Nissan Sentra 4 door	1992-1994	92	1995-1997	66	26
Honda Civic 2 door	1993-1995	86	1996-1998	71	15
Honda Accord 2 door	1995-1997	72	1998-2000	63	9
Honda Accord 4 door	1995-1997	77	1998-2000	59	18
Subaru Legacy 4 door	1997-1999	68	2000-2002	48	20
Subaru Impreza 4 door	1999-2001	72	2002	45	27
Toyota Corolla/Geo Prizm (4 door)	1994-1996	91	1997-1999	66	25
Mitsubishi Eclipse 2 door	1997-1999	89	2000-2002	52	37

REGISTRATION-WEIGHTED AVERAGE TTI(d) IMPROVEMENT: 23

The maximum range permitted was 6 model years: the first 3 years with low TTI(d) vs. the last 3 years before the improvement. The purpose of limiting the model-year range is to prevent excess disparity in vehicle ages between the low-TTI(d) and high-TTI(d) cars. As cars get substantially older, the type and average severity of their crashes may change, possibly biasing the analyses. On the other hand, making the range even shorter might not provide enough data for statistically meaningful results.[62] However, the "before" range has to be cut off sooner if a make-model was introduced (Dodge Intrepid) or substantially redesigned (Chevrolet Cavalier) less than 3 years before the TTI(d) improvement. Similarly, the low-TTI(d) range has to be cut off sooner if a make-model was discontinued, substantially redesigned (Pontiac Grand Am) or received side air bags as standard equipment less than 3 years after the TTI(d) improvement, or when the range reaches 2002, the last model year with extensive FARS data (Subaru Impreza). (Cars with optional side air bags, or that cannot be determined if they have side air bags, are excluded from the analyses.)

Table 2-2 lists the 12 make-model groups in Category 2. These make-models had little or no change in TTI(d) when they were originally certified to FMVSS 214, nor did they later experience any substantial improvement (without side air bags) that might have moved them to Category 1. Category 2 is one of the most important informal "control groups" for checking the analysis results: here, side impact fatality rates ought to be about the same before and after FMVSS 214 certification.

IR's and/or test results, preferably both, can be a basis for assigning make-models to Category 2. Lincoln Town Car, Mercury Grand Marquis, Chevrolet Caprice and Chevrolet Camaro were not redesigned in the year they were certified to FMVSS 214, and the IR's indicate no change at all, or very little change for the purpose of meeting FMVSS 214. Furthermore, except for Camaro, we have test results for pre-standard vehicles and they are as good as, or better than the certified vehicles. For the other seven make-model groups, we have test results before and after certification, and they show at most a small TTI(d) decrease, 5 units or less, for the certified cars. For the Ford Escort 4-door, Ford Probe, Buick LeSabre, Chevrolet Cavalier 4-door, Chevrolet Lumina 4-door and Pontiac Grand Am 4-door, the test results are essentially corroborated by IR's that only mention padding and/or minor structural enhancements. By contrast, the IR's for the Saturn 4-door and Honda Civic 4-door describe major structural reinforcements in 1996-1997, even though our tests do not show any TTI(d) improvement from 1995. As stated above, test results can vary, and an exceptionally low reading followed by an exceptionally high one could have masked a substantial TTI(d) improvement.

For the 13 remaining make-model groups, NHTSA has no test results before FMVSS 214 certification. Table 2-3 lists the 5 groups assigned to Category 3 because the IR indicates, or at least seems to indicate that the cars received major structural improvements plus padding in the year they certified to FMVSS 214. We must resist the temptation to add them to Category 1, because without actual test scores for pre-standard cars, there is no hard evidence that TTI(d) substantially decreased (or that we correctly interpreted the IR's). But they will remain separate from the other 8 make-model groups without pre-standard test results, the Category 4 cars shown in Table 2-4.

---

[62] Kahane, C.J., *Fatality Reduction by Air Bags*, NHTSA Technical Report No. DOT HS 808 470, Washington, 1996, p. 9.

TABLE 2-2: CATEGORY 2 MAKE-MODEL GROUPS WITH NO TTI(d) CHANGE OR LIMITED CHANGE

Make-Model(s)	"Before" Model Years	TTI(d)	"After" Model Years	TTI(d)	Δ TTI(d)
Ford Escort 4 door	1994-1996	67	1997-1999	65	2
Ford Probe/Mazda MX-6	1994-1996	83	1997	80	3
Lincoln Town Car (4 door)	1991-1993	40 (?)	1994-1996	63	None[63]
Mercury Grand Marquis/Crown Victoria (4 door)	1991-1993	41 (?)	1994-1996	51	None[64]
Buick LeSabre/Olds 88/Pontiac Bonneville (4 door)	1994-1996	72	1997-1999	67	5
Chevrolet Caprice/Roadmaster 4 door sedan	1991-1993	50 (?)	1994-1996	56	None[65]
Chevrolet Camaro/Pontiac Firebird 2 door coupe	1993-1994	Unk.	1995-1996	77	None[66]
Chevrolet Cavalier/Pontiac Sunfire 4 door	1995-1996	82[67]	1997-1999	78	4
Chevrolet Lumina 4 door	1992-1994	66	1995-1997	62	4
Pontiac Grand Am/Skylark/Achieva 4 door	1994-1996	64	1997-1998	72	- 8 (?)
Saturn 4 door	1993-1995	69	1996-1998	69	None
Honda Civic 4 door	1993-1995	54	1996-1998	61	- 7 (?)

44

[63] Manufacturer responded to NHTSA Information Request, stated there was little or no vehicle modification for the purpose of certifying to FMVSS 214.
[64] Manufacturer, in response to NHTSA Information Request, did not report any vehicle modifications for the purpose of certifying to FMVSS 214.
[65] Manufacturer responded to NHTSA Information Request, stated the vehicle was unchanged upon certifying to FMVSS 214.
[66] Manufacturer responded to NHTSA Information Request, stated there was little or no vehicle modification for the purpose of certifying to FMVSS 214.
[67] The only pre-standard Cavaliers tested were a 1986 and a 1987 model, and their TTI(d) averaged 82. The average TTI(d) for 214-certified 1997-1998 Cavaliers was 78. We may infer that TTI(d) probably remained close to 82 throughout 1986-1996. While it is not impossible that TTI(d) became much worse from 1986-1987 to 1993-1996 and then greatly improved in 1997 – or that it became much better from 1986-1987 to 1993-1996 and then greatly deteriorated upon certification in 1997 – both are quite unlikely.

TABLE 2-3: CATEGORY 3 MAKE-MODEL GROUPS THAT MAY HAVE RECEIVED MAJOR NEW STRUCTURE
UPON FMVSS 214 CERTIFICATION, TTI(d) CHANGE UNKNOWN

| Make-Model(s) | "Before" | | "After" | | Δ |
	Model Years	TTI(d)	Model Years	TTI(d)	TTI(d)
Mazda Protégé (4 door)	1992-1994	Unk.	1995-1997	64	Unk.
Toyota Celica (2 door)	1994-1995	Unk.	1996-1998	59	Unk.
Toyota Tercel 4 door	1992-1994	Unk.	1995-1997	63	Unk.
Toyota Camry 4 door	1992-1993	Unk.	1994-1996	65	Unk.
Mitsubishi Galant (4 door)	1991-1993	Unk.	1994-1996	69	Unk.

TABLE 2-4: CATEGORY 4 MAKE-MODEL GROUPS WITH UNKNOWN TTI(d) CHANGE UPON FMVSS 214 CERTIFICATION, NO EVIDENCE OF MAJOR NEW STRUCTURE

Make-Model(s)	"Before"		"After"		Δ TTI(d)
	Model Years	TTI(d)	Model Years	TTI(d)	
Ford Thunderbird/Mercury Cougar (2 door)	1992-1994	Unk.	1995-1997	70	Unk.
Buick Park Avenue/Olds 98 (4 door)	1994-1996	Unk.	1997-1999	62	Unk.
Cadillac DeVille 4 door	1991-1993	Unk.	1994-1996	54	Unk.
Geo Metro/Suzuki Swift 2 door	1992-1994	Unk.	1995-1997	80	Unk.
Saturn 2 door	1994-1996	Unk.	1997-1999	74	Unk.
Nissan Maxima (4 door)	1992-1994	Unk.	1995-1997	58	Unk.
Nissan Altima (4 door)	1994-1996	Unk.	1997-1999	69	Unk.
Mazda 626 (4 door)	1993-1995	Unk.	1996-1997	66	Unk.

In Category 4, not only is the TTI(d) decrease upon certification unknown, but also the IR information is unavailable or ambiguous. There is no IR for Buick Park Avenue. The IR's for Ford Thunderbird, Saturn 2-door, Nissan Altima and Mazda 626 do not clearly indicate if structural modifications were substantial or minor. For Geo Metro and Nissan Maxima, the IR does not spell out if the described changes were implemented upon certification or in a later year. The IR says Cadillac DeVille was "not modified" to meet FMVSS 214, but because it was redesigned in the year of certification, it is not the same vehicle as the preceding year.

2.3 Fatalities per 1,000 nearside occupants in multivehicle side impacts

A basic analysis of the effect of TTI(d) improvements is to compute fatality rates per 1,000 nearside occupants in multivehicle side impacts, and to compare these rates in the Category 1 make-models before and after the substantial TTI(d) improvement. Nearside occupants in multivehicle side impacts are of particular interest because they are the occupants and impacts that most closely resemble the FMVSS 214 test configuration. But there is another reason for concentrating on them: other occupant protection measures, specifically safety belts and frontal air bags have little effect for nearside occupants in side impacts by a vehicle.[68] It makes little difference if the newer, low-TTI(d) cars have higher belt use or more frontal air bags, and it is not necessary to control for those factors.

The fatality rates are generated from two data files. The numerator – the number of fatalities – derives from the Fatality Analysis Reporting System (FARS), a census of the nation's fatal crashes, but lacking information on crashes where nobody died. The denominator – the number of nearside front-seat occupants of cars that were struck in the side by another vehicle – derives from the General Estimates System (GES) of the National Automotive Sampling System (NASS). It is a probability sample of the nation's crash involvements, and when GES cases are weighted by the inverse sampling fractions they generate unbiased estimates of national totals. However, the number of fatality cases within GES itself is small, raising the sampling error of fatality rates based on GES alone. Intuitively, rates obtained by using the copious FARS census data for the numerator, and GES data only for the denominator ought to be much more accurate.[69]

The analysis is based on calendar year 1993-2005 FARS and GES data. It includes the Category 1 make-models for the ranges of model years shown in Table 2-1b – i.e., up to 3 model years before and after the transition to a lower TTI(d). The make-model of a car can be identified in GES by decoding the VIN and/or from GES' own make-model codes. Approximately 14 percent of GES cases have missing VIN and make-model codes, but since that percentage is the same just before and just after the TTI(d) improvement (i.e., does not vary with vehicle age or by calendar year) it should not influence the effectiveness estimate. Other analysis variables, such as the VIN on FARS and the impact location on both files have less than 3 percent missing data.

---

[68] As discussed in Section 1.1, NHTSA found only a non-significant 5 percent reduction in fatality risk for belt use in nearside multivehicle crashes, as compared to statistically significant reductions of 21 percent in nearside impacts with fixed objects and 39 percent in farside impacts. Likewise, air bags are of little value except in the small proportion of multivehicle crashes that include a substantial oblique-frontal force component.

[69] For examples of analyses combining FARS and GES see Joksch, H., *Vehicle Design versus Aggressivity*, NHTSA Technical Report No. DOT HS 809 184, Washington, 2000.

In FARS data, "nearside occupants" include drivers when the principal impact point is 8, 9 or 10:00 and right-front passengers when the principal impact is 2, 3 or 4:00.[70] "Multivehicle crashes" are those that involve two or more vehicles and the first harmful event was a "collision of motor vehicles in transport."

Likewise, in GES data, "nearside occupants" include drivers when the principal impact point is on the left side and right-front passengers when the principal impact is on the right side. "Multivehicle crashes" are those that involve two or more vehicles and the first harmful event was a "collision of motor vehicles in transport."

Confounding of "towaway" with side structure improvement Past analyses of crash data have often computed fatality or injury rates per 1,000 occupants involved in towaway crashes, because "towaway" seems an objective and invariant threshold of crash severity, not dependent on State or local variation of what crashes should be and, in fact, are reported. That criterion cannot be used here because improvements to side structures, in addition to protecting occupants, could make a car more damage-resistant. The car might be driven away when, without the improvements, it might have been towed after the same impact. In the 15 make-model groups of Category 1, 37.4 percent of side impacts on GES were towaways in the model years before the TTI(d) reduction, but only 35.6 percent in cars of the model years after the reduction. By contrast, in frontal crashes (where the side structure would have little effect on damageability), 43.4 percent were towed away before the TTI(d) reduction and a nearly identical 43.5 percent afterwards. Likewise, in the Category 2 make-models, where TTI(d) stayed about the same before and after FMVSS 214, the proportion of towaways in side impacts of pre-standard cars, 33.4 percent was actually lower than in the post-standard cars, 34.9 percent.

Limiting the analysis to towaway crashes would have biased the results against TTI(d) improvement, because the cars with lower TTI(d) had, on the average, more severe towaway crashes, and consequently higher fatality rates per 1,000 towaway crashes.

Basic analysis of fatality rates Table 2-5 shows a 21 percent reduction in the fatality rate for the 15 Category 1 make-models that substantially improved TTI(d). In the model years before the TTI(d) improvement, there were 1,452 fatalities among the estimated 458,872 nearside occupants in side impacts by another vehicle, a fatality rate of 3.16 per thousand occupants. In these same make-models, after the TTI(d) improvement, the fatality rate was 2.49 (1,096 fatalities per 440,109 occupants).

70

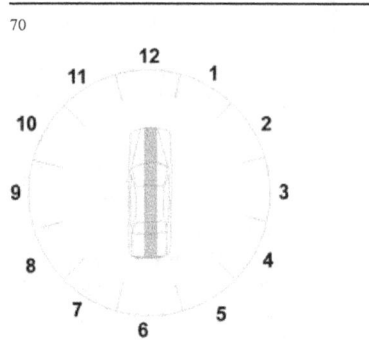

TABLE 2-5

MAKE-MODELS WITH SUBSTANTIAL TTI(d) IMPROVEMENTS, FATALITIES PER
1,000 NEARSIDE OCCUPANTS IN SIDE IMPACTS BY ANOTHER VEHICLE
(1993-2005 FARS and GES data)

	Nearside Fatalities	Nearside Occupants	Fatality Rate	Fatality Reduction
Before TTI(d) improvement	1,452	458,872	3.16	
After TTI(d) improvement	1,096	440,109	2.49	21 %

The 15 make-models in Table 2-1b include six "second generation" TTI(d) improvements after the 1994-1997 phase-in of FMVSS 214, stretching out to model year 2002 in some cases. As discussed in Section 1.4, there may be overlap with the installation of energy-absorbing materials for head-impact protection in response to FMVSS 201 (1999-2003 phase-in). But, statistically, the overlap is minimal: in Table 2-5, 100 percent of the cars before the TTI(d) improvement and 89 percent after the TTI(d) improvement are pre-FMVSS 201. Of the remaining 11 percent, over two-thirds are Honda Accords (that certified to FMVSS 201 in 1999); NHTSA's cost analysis of FMVSS 201 found relatively little energy-absorbing material added in the Accord, relative to other models studied.[71] Thus, at most 3½ percent of the "after" sample in Table 2-5 may have substantial modifications related to FMVSS 201.

Statistical significance of the fatality reduction can be tested and confidence bounds computed by treating the effectiveness as a ratio of ratios of FARS and GES statistics. FARS is a census, but for the purpose of analyzing uncertainty of effectiveness estimates, NHTSA treats it as a simple random sample from an infinitely large population.[72] GES is a cluster sample of primary sampling units (PSU) that are groups of counties, with a census of crashes within the PSUs. The FARS data were split up into 10 systematic random subsamples, numbered 0 to 9, based on the last digit of the case identification number ST_CASE. The GES **PSUs** were randomly split into 10 subgroups of approximately equal size, as follows:

- The PSUs were randomly ordered by issuing each PSU a new PSU number with a SAS random-number generator, and listed in this new order.

- The number of weighted vehicle cases (all vehicle types, all crash types) during 1993-2005 was ascertained for each PSU and cumulated down the list.

- The list was parsed into consecutive (by the new PSU numbers) groups of PSUs, each containing about 10 percent of the cases, and these subgroups were numbered 0 to 9.

---

[71] Ludtke, N.F., Osen, W., Gladstone, R. and Lieberman, W., *Perform Cost and Weight Analysis, Non Air Bag Head Protection Systems, FMVSS 201*, NHTSA Technical Report No. DOT HS 809 810, Washington, 2003, pp. 2-10 and 3-9 – 3-12.

[72] The hypothetical "infinite population" would be the endless repetition of the driving environment in the United States. For example, if there were 1,000 fatalities of a specific type in the United States in 2004, that number could be somewhat higher or lower the next year even if the driving environment were to stay essentially the same (Poisson variation, in this case).

Effectiveness, more specifically

$$\log r = \log [(\text{fatals}_{\text{post}} / \text{occs}_{\text{post}}) / (\text{fatals}_{\text{pre}} / \text{occs}_{\text{pre}})]$$

is calculated using only the data from FARS subgroup 1 and GES subgroup 1, then recalculated using subgroups 2, and so on, to obtain ten estimates, each based on about 1/10 of the data in Table 2-5. The standard deviation of these ten estimates is .1646. The standard deviation of the corresponding estimate in the full dataset (10 times as many data) is $.1646 / \sqrt{10} = .0521$. Based on Table 2-5, the estimate for the full dataset is

$$\log [(1,096 / 440,109) / (1,452 / 458,872)] = -.2395$$

Because $t = -.2395/.0521 = -4.60$ is more negative than -1.833, the 5th percentile of a t distribution with 9 degrees of freedom, the fatality reduction is statistically significant at the one-sided .05 level. (In fact, because -4.60 is more negative than -3.25, the 0.5^{th} percentile of a t distribution with 9 degrees of freedom, the reduction is significant at the two-sided .01 level.)

The 90 percent confidence bounds[73] for log r are $-.2395 \pm 1.833 \times .0521 = (-.3350, -.1440)$. The 90 percent confidence bounds for fatality-reducing effectiveness $(1 - \exp[\log r])$ range from **13 to 28 percent**.

However, given that these fatality rates are based on 15 make-models, a more stringent test of significance is to compare results for the individual make-models. There was a fatality reduction in 11 of the 15 make-models after the TTI(d) improvement. If a balanced coin is flipped 15 times, there is a 5.9 percent probability that it will land heads 11 times or more (exact binomial test). In other words, the hypothesis that fatality rates were the same before and after TTI(d) improvement can be rejected at the .10 level but not quite at the .05 level; by this test the observed fatality reduction is statistically significant only at the .10 level.

Fatality rates in Category 2-4 make-models By contrast, for the control group of 12 Category 2 make-models whose TTI(d) was essentially unchanged upon FMVSS 214 certification, Table 2-6 shows a 7 percent increase in the fatality rate after certification.

TABLE 2-6

MAKE-MODELS <u>WITHOUT</u> TTI(d) IMPROVEMENTS, FATALITIES PER 1,000
NEARSIDE OCCUPANTS IN SIDE IMPACTS BY ANOTHER VEHICLE
(1993-2005 FARS and GES data)

	Nearside Fatalities	Nearside Occupants	Fatality Rate	Fatality Reduction
Before 214 certification	1,486	438,236	3.39	
After 214 certification	1,312	360,073	3.64	− 7 %

---

[73] -1.833 and +1.833 are the 5th and 95th percentiles of a t-distribution with 9 degrees of freedom.

The 7 percent increase is not statistically significant (t = 1.32 in the analysis of sampling error similar to the procedure for Table 2-5). Five of these make-models had lower fatality rates after certification, and seven had higher rates.

Table 2-7 displays the fatality rates of the two remaining categories of make-models, the ones where it is unknown how much TTI(d) changed, if at all, upon FMVSS 214 certification. In the five make-models of Category 3, the IR's suggest possible major structural modifications upon FMVSS 214 certification, but NHTSA has no test results for the immediate pre-standard models and does not know how much TTI(d) changed, if at all. The N of crash cases is only about ¼ as large as in Categories 1 or 2. Fatality rates are 5 percent higher after certification; the increase is not statistically significant. Category 4, where TTI(d) change is unknown and NHTSA has little evidence of major structural modification, has fewer than half the cases of Categories 1 or 2. Fatality risk is a non-significant 2 percent lower after FMVSS 214 certification.

TABLE 2-7

MAKE-MODELS WITH <u>UNKNOWN</u> TTI(d) IMPROVEMENTS, FATALITIES PER 1,000
NEARSIDE OCCUPANTS IN SIDE IMPACTS BY ANOTHER VEHICLE
(1993-2005 FARS and GES data)

	Nearside Fatalities	Nearside Occupants	Fatality Rate	Fatality Reduction
Category 3: perhaps major new structure upon 214 certification, TTI(d) change unknown				
Before 214 certification	278	108,706	2.56	
After 214 certification	303	112,439	2.70	– 5 %
Category 4: no evidence of major new structure upon 214 certification, TTI(d) change unknown				
Before 214 certification	580	212,955	2.72	
After 214 certification	414	155,711	2.66	2 %

Thus, only the Category 1 models consistently showed a reduction of nearside impact fatalities per 1,000 exposed occupants. The fatality reduction was 21 percent following substantial TTI(d) improvements averaging 23 units. Neither the control-group make-models with unchanged TTI(d) (Category 2), nor the make-models with unknown TTI(d) history (Categories 3 and 4) experienced a significant fatality reduction upon FMVSS 214 certification.

Another control: fatality rates in frontal crashes Corresponding fatality rates may be calculated for multivehicle frontal impacts to check if the effects seen in Tables 2-5 and 2-6 are unique to nearside impacts (where TTI(d) improvements ought to have an effect, but not in frontals). An initial problem with frontals, unlike nearside impacts, is that safety belts and frontal air bags are quite effective. Cars have higher belt use and may be more likely to be equipped with frontal air bags in later model years – i.e., subsequent to the TTI(d) improvement. A remedy for this possible bias is to inflate the observed number of FARS fatality cases of occupants protected by safety belts and/or air bags by the inverse of the effectiveness, using the method of NHTSA's evaluation of lives saved by the FMVSS.[74] For example, if safety belts reduced fatality risk by 45 percent and air bags by 30 percent in a certain type of frontal crash, each fatality case of a belted occupant in a car equipped with frontal air bags in that type of crash would be given a weight of

$$1/ [(1 - .45) (1 - .3)] = 2.6 \text{ fatalities}$$

The sum of these weights is an estimate of the number of fatalities that would have occurred if none of the occupants had been protected by safety belts or frontal air bags. Frontal crashes are defined by the principal impact region (11, 12 or 1:00 on FARS, "front" on GES). "Multivehicle crashes" are those that involve two or more vehicles and the first harmful event was a "collision of motor vehicles in transport." The analysis is limited to drivers and right-front passengers.

Table 2-8 shows the frontal fatality rates in the 15 Category 1 make-models that substantially improved TTI(d). In the model years before the TTI(d) improvement, there were 3,404 adjusted frontal fatalities among 1,248,188 occupants in frontal multivehicle impacts, a fatality rate of 2.73 per thousand occupants. After the TTI(d) improvement, the fatality rate was 2.63. That is a non-significant 3 percent reduction. It is negligible compared to the 21 percent reduction in the nearside impacts.

TABLE 2-8

MAKE-MODELS <u>WITH</u> SUBSTANTIAL TTI(d) IMPROVEMENTS, FATALITIES PER 1,000 OCCUPANTS IN FRONTAL IMPACTS BY ANOTHER VEHICLE
(1993-2005 FARS and GES data; fatalities adjusted upward for belt use and frontal air bags)

	Frontal Fatalities	Exposed Occupants	Fatality Rate	Fatality Reduction
Before TTI(d) improvement	3,404	1,248,188	2.73	
After TTI(d) improvement	2,706	1,029,050	2.63	3 %

In the control group of 12 Category 2 make-models whose TTI(d) was essentially unchanged upon FMVSS 214 certification, Table 2-6 showed a non-significant 7 percent increase in the nearside fatality rate after certification. Table 2-9 shows an identical 7 percent increase in the frontal fatality rate, after adjusting for safety belt use and frontal air bags. Nearside fatality rates neither increased nor decreased relative to frontal fatality rates.

---

[74] Kahane, C.J., *Lives Saved by the Federal Motor Vehicle Safety Standards and Other Vehicle Safety Technologies, 1960-2002*, NHTSA Technical Report No. DOT HS 809 833, Washington, 2004, pp. 173-182, 309-312 and 316-317.

TABLE 2-9

MAKE-MODELS <u>WITHOUT</u> TTI(d) IMPROVEMENTS, FATALITIES PER 1,000
OCCUPANTS IN FRONTAL IMPACTS BY ANOTHER VEHICLE
(1993-2005 FARS and GES data; fatalities adjusted upward for belt use and frontal air bags)

	Frontal Fatalities	Exposed Occupants	Fatality Rate	Fatality Reduction
Before 214 certification	3,439	994,956	3.46	
After 214 certification	2,788	754,683	3.69	– 7 %

A caveat – absolute fatality rates The aggregate nearside fatality rates for the Category 2 cars with unchanged TTI(d), 3.39 before FMVSS 214 certification and 3.64 after (see Table 2-6) are higher than the corresponding rates for Categories 1, 3 and 4, ranging from 2.49 up to 3.16 (see Tables 2-5 and 2-7). Dainius Dalmotas, in his review of this report, pointed out that Category 2 includes some make-models with excellent TTI(d) scores before as well as after certification, and might be expected to have low, not high fatality rates.

It is true that a strong relationship between TTI(d) and absolute fatality rates, by make-model, would have made the optimum case for the effect of TTI(d) reductions. But such a relationship may not exist because there are many other factors that cause fatality rates per 1,000 reported crashes to vary widely from model to model. Specifically, Category 2 includes:

- Several full-sized models favored by older drivers, who are especially vulnerable to fatal injury, given a crash.

- Sporty coupes as well as models widely used as police cars, prone to high-severity crashes.

- Generally earlier model years than Category 1 – i.e., somewhat older cars, with more fatalities per 1,000 reported crashes than new cars (lower reporting of minor crashes).

Thus, even in frontal crashes, where TTI(d) is unlikely to be a factor, the absolute fatality rates for Category 2 (Table 2-9) are similarly elevated relative to Category 1 (Table 2-8).

Fatality reduction vs. TTI(d) improvement by make-model Nearside fatality rates can vary for many reasons between make-models. But within a given make-model, driver characteristics and use patterns tend to be relatively stable across model years. If fatality rates change over time, it may well be attributable to vehicle design. Specifically, it is reasonable to expect that, the larger the TTI(d) improvement for a make-model, the greater the nearside fatality reduction for that model, post- vs. pre-improvement.[75]

The 27 make-models of Categories 1 and 2 are the only make-models where the TTI(d) improvement is quantitatively known – as shown in Tables 2-1b and 2-2 in the "Δ TTI(d)"

---

[75] This analysis was added in response to Dalmotas' recommendation to investigate parametric relationships between TTI(d) and fatality risk, in addition to comparison of aggregate fatality rates as in Table 2-5.

columns (positive number = improvement). They are the 27 data points for a parametric analysis. Based on the same FARS and GES data used in Tables 2-5 and 2-6, the change in fatality risk is expressed as the log of the ratio of the nearside multivehicle fatality rate after the TTI(d) improvement to the rate before the improvement (negative number = fatality reduction). Each data point is weighted by the number of fatality cases for that model before the improvement, or the number after, whichever is smaller.

The correlation between the amount of TTI(d) improvement and the amount of fatality reduction is $-.384$, and it is statistically significant at the two-sided .05 level. A weighted regression produces a statistically significant coefficient of $-.00803$ for Δ TTI(d). In other words, a one-unit reduction in TTI(d) is associated with a 0.803 percent reduction in the fatality rate.[76] The 23-unit reduction in TTI(d) for the make-models in Category 1 would scale out to a

$$1 - (1 - .00803)^{23} = 17 \text{ percent fatality reduction}$$

Effect of the type of striking vehicle Approximately 90 percent of FARS and GES nearside impacts in multivehicle crashes are two-vehicle crashes where the body type of the striking vehicle is known, and is one of the following:

- A passenger car.

- A light truck or van (LTV), including pickup trucks, SUVs, minivans and full-size vans up to 10,000 pounds Gross Vehicle Weight Rating (GVWR).

- A tractor-trailer, heavy truck or bus over 10,000 pounds GVWR.

Table 2-10 analyzes nearside fatality rates in the 15 Category 1 make-models that substantially improved TTI(d). When the striking vehicle was a passenger car, the nearside fatality rate in the struck car was reduced 22 percent after the TTI(d) improvement. That is similar to the effect in all nearside impacts (21 percent; see Table 2-5). But when the striking vehicle is an LTV, the observed fatality reduction in the struck vehicle after TTI(d) improvement is slightly, but not significantly larger: 31 percent.[77] In other words, the fatality reduction is substantial, and about the same, regardless whether the striking vehicle is a car or an LTV.[78] When a heavy truck is the striking vehicle, the fatality rate is 1 percent higher in the cars with improved TTI(d); this last statistic, however, is based on a much smaller N of cases than the preceding two. (By contrast, in the 12 control-group, unchanged-TTI(d) make-models of Category 2, each of these three fatality rates stayed about the same before and after FMVSS 214 certification.)

---

[76] Based on the CORR and GLM procedures in SAS, *SAS/STAT® User's Guide*, Vol. 1, Version 6, 4th Ed., SAS Institute, Cary, NC, 1990.

[77] Given that the confidence bounds on the estimate in Table 2-5 are ± 7 or 8 percent, a 9 percentage-point difference in two effectiveness estimates each based on less than half the data from Table 2-5 is not significant.

[78] In Table 2-10 the absolute fatality rates are higher when the striking vehicle is an LTV than when it is a car. That could reflect, among other things, higher aggressiveness of LTVs (LTV-car incompatibility) and/or a greater proportion of LTVs in less urbanized areas, where crashes tend to be more severe.

TABLE 2-10

MAKE-MODELS <u>WITH</u> SUBSTANTIAL TTI(d) IMPROVEMENTS
FATALITIES PER 1,000 NEARSIDE OCCUPANTS IN SIDE IMPACTS
BY <u>TYPE OF STRIKING VEHICLE</u>
(1993-2005 FARS and GES data, 2-vehicle crashes)

	Nearside Fatalities	Nearside Occupants	Fatality Rate	Fatality Reduction
Striking vehicle is a passenger car				
Before TTI(d) improvement	412	256,081	1.61	
After TTI(d) improvement	294	234,121	1.26	22 %
Striking vehicle is an LTV				
Before TTI(d) improvement	698	135,766	5.14	
After TTI(d) improvement	502	142,416	3.53	31 %
Striking vehicle is a heavy truck				
Before TTI(d) improvement	170	28,011	6.07	
After TTI(d) improvement	163	26,561	6.14	− 1 %

Ejected vs. non-ejected fatalities How much, if any of the overall fatality reduction in Category 1 make-models (Table 2-5) is due to fewer occupants being ejected from their cars? Specifically, FMVSS 214 includes a separate door-retention requirement (see Section 1.3). Table 2-11 shows, however, that well under 10 percent of the nearside fatalities in multivehicle crashes were ejected (totally or partially). The impacts are quite dangerous even for occupants who remain within the vehicle, and the probability of ejection has been reduced in recent years by higher belt use. Thus, if TTI(d) reductions have a substantial overall effect, most of it will be on the non-ejected occupants.

Table 2-11 shows that the fatality risk of non-ejected occupants was 21 percent lower, after the TTI(d) improvement, in the Category 1 make-models. That is the same as the overall reduction in Table 2-5. Moreover, the Category 1 make-models are the only group where the fatality risk of non-ejected occupants was lower after TTI(d) improvement/FMVSS 214 certification. In Category 1, the rate of ejected occupant fatalities was reduced by 24 percent, almost the same reduction as for the non-ejected fatalities. Moreover, there was a similar reduction of ejected fatalities in each of the other groups. Perhaps some of that effect is due to higher belt use in the newer vehicles. In any case, the contribution of ejection-reduction to the overall effect in Category 1 is small, because so few of the fatalities were ejected.

TABLE 2-11

EJECTED VS. NON-EJECTED FATALITIES PER 1,000 NEARSIDE OCCUPANTS IN SIDE IMPACTS BY ANOTHER VEHICLE
(1993-2005 FARS and GES data)

	Nearside Occupants	Non-Ejected Fatalities			Ejected Fatalities		
		N	Rate	Red.	N	Rate	Red.
Category 1: make-models with substantial TTI(d) improvements							
Before TTI(d) improvement	458,872	1,364	2.97		88	.192	
After TTI(d) improvement	440,109	1,032	2.34	21 %	64	.145	24 %
Category 2: make-models without TTI(d) improvements							
Before 214 certification	438,236	1,406	3.21		80	.183	
After 214 certification	360,073	1,261	3.50	− 9 %	51	.142	22 %
Category 3: perhaps major new structure upon 214 certification, TTI(d) change unknown							
Before 214 certification	108,706	259	2.38		19	.175	
After 214 certification	112,439	286	2.54	− 7 %	17	.151	14 %
Category 4: no evidence of major new structure upon 214 certification, TTI(d) change unknown							
Before 214 certification	212,955	536	2.52		44	.207	
After 214 certification	155,711	394	2.53	− 1 %	20	.128	38 %

56

Logistic regression, an alternative to the basic analysis TTI(d) improvement coincided with an overall redesign in 11 of the 15 Category 1 make-models, and when cars are redesigned, the types of people who drive them sometimes changes. As a consequence, the severity of their crashes may also change. With logistic regression and these data, we can control for the two most important quantifiable human parameters – drivers' age and gender – plus three variables that are strongly associated with crash severity and are defined the same way on FARS and GES: the speed limit, the time of the day, and the type of striking vehicle.

The data points in the logistic regression are the FARS occupant fatality cases from Table 2-5, each given a weight factor of 1, and the **nonfatal** GES occupant cases from Table 2-5, each given a weight factor equal to the inverse of the sampling fraction. Cases with unreported age or gender are excluded; so are right-front passengers younger than 12 years. The dependent variable, FATAL equals 1 if the occupant was a fatality, 2 if not. The key independent variable, TTI_IMPR is the TTI(d) status: 0 before the improvement vs. 1 after the improvement. The other independent variables are:

- The occupant's age, expressed as AGEGE18, the age minus 18 (but set to zero if the occupant was 12-18 years old). This recognizes that fatality risk given a similar physical insult rises steadily from age approximately 18 onward, but not from 12 to 18.[79]

- Gender, expressed as FEMALE (= 1 for females, 0 for males)

- Seat position, expressed as RF (= 1 for right-front passengers, 0 for drivers)

- Vehicle age (VEHAGE)

- Speed limit, expressed as SPDLIM55 (= 1 if it is 55+, 0 if it is 50 or less; when the various vehicles are on different roads, this is the highest speed limit for any of the roads)

- Time of day, expressed as NITE (= 1 if 7:00 p.m. – 5:59 a.m., 0 if 6:00 a.m. – 6:59 p.m.)

- The type of striking vehicle (car, LTV, heavy truck, or other/unknown/3+ vehicle crash), expressed as three dichotomous variables OV_LTV, OV_HVYTK, OV_N_AB [none of the above]; all are set to zero if the other vehicle was a passenger car

The initial regression is based on 2,518 FARS fatality cases and 850,307 weighted (7,553 unweighted) GES nonfatal cases. The regression coefficients are:

---

[79] Evans, L., *Traffic Safety and the Driver,* Van Nostrand Reinhold, New York, 1991, pp. 25-28.

NEARSIDE FATALITY RISK IN SIDE IMPACTS BY ANOTHER VEHICLE

	Coefficient	*"Wald Chi-Square"*	*"P <"*
TTI_IMPR	**- .268**	*42.7*	*.0001*
AGEGE18	.0354	*1308.*	*.0001*
FEMALE	- .196	*23.0*	*.0001*
RF	1.046	*560.*	*.0001*
VEHAGE	.043	*41.3*	*.0001*
SPDLIM55	1.531	*1339.*	*.0001*
NITE	.418	*78.2*	*.0001*
OV_LTV	1.013	*443.*	*.0001*
OV_HVYTK	.817	*136.*	*.0001*
OV_N_AB	1.426	*420.*	*.0001*
INTERCEPT	- 7.947	*13737.*	*.0001*

The above chi-square and P values should not be taken literally, since they are calculated as if the weighted N of nonfatal cases (850,307) were actually separate observations from a simple random sample. Instead, they are only 7,553 observations from the GES cluster sample. Whatever has $P < .05$ in the above table is not necessarily significant, but whatever has $P \geq .05$ is definitely non-significant.

The coefficient for TTI_IMPR is -.268. In other words, TTI(d) improvement is associated with a $1 - \exp(-.268) = 24$ percent reduction in fatality risk. That is slightly higher than the 21 percent reduction based on the simple comparison of fatality rates in Table 2-5. Controlling for factors such as occupant age/gender or crash type/location did not substantially change the net effectiveness.

A second logistic regression adds 14 dichotomous variables, one for each of the make-model groups except Taurus (and when all 14 of these variables are zero, the case vehicle is a Taurus). The purpose of these variables is to control for any imbalance in the data due to a relative excess of one (or several) make-model(s) among the pre-improvement cars. With the additional control variables, the coefficient for TTI_IMPR drops a little bit to -.247, equivalent to a $1 - \exp(-.247) = 22$ percent reduction in fatality risk. That is almost identical to the 21 percent reduction based on the simple comparison of fatality rates.

Logistic regression can also be used to analyze the "effect" of TTI(d) improvement of fatality risk in frontal crashes. The data points are the FARS occupant fatality cases from Table 2-8, inflated (i.e., given a weight factor greater than 1), if necessary, to adjust for safety belt use and frontal air bags, and the nonfatal GES occupant cases from Table 2-8, each given a weight factor equal to the inverse of the sampling fraction.[80] The regression coefficients are:

---

[80] Cases with unreported age or gender are excluded; so are right-front passengers younger than 12 years.

FATALITY RISK IN FRONTAL IMPACTS BY ANOTHER VEHICLE

	Coefficient	*"Wald Chi-Square"*	*"P <"*
TTI_IMPR	**- .002**	*.003*	*.95*
AGEGE18	.0373	*3118.*	*.0001*
FEMALE	- .300	*127.*	*.0001*
RF	.527	*259.*	*.0001*
VEHAGE	.062	*193.*	*.0001*
SPDLIM55	2.201	*6490.*	*.0001*
NITE	1.066	*1368.*	*.0001*
OV_LTV	.605	*328.*	*.0001*
OV_HVYTK	2.441	*3790.*	*.0001*
OV_N_AB	2.028	*2657.*	*.0001*
INTERCEPT	- 8.525	*33330.*	*.0001*

The coefficient for TTI_IMPR is -.0002 equivalent to an exp(-.002) – 1 = 0.2 percent reduction in fatality risk – i.e., no change. That is similar, but even more neutral than the 3 percent reduction in Table 2-8's simple comparison of fatality rates. The newer/redesigned models had slightly less severe frontal crashes than their predecessors. Controlling for these crash-severity variables eliminates even the small, non-significant frontal fatality reduction "attributed" to TTI(d) improvement in the simple comparison.

In a second regression with the additional 14 make-model variables, the coefficient for TTI_IMPR in frontal crashes is +.029, equivalent to a non-significant 3 percent increase in the fatality risk.

The regression results corroborate the nearside fatality reduction after TTI(d) improvement in Table 2-5 and the absence of a significant effect in frontal crashes, in Table 2-8.

2.4   Nearside fatalities in multivehicle crashes relative to non-occupant fatalities

Another approach is to consider non-occupant fatalities a control group and to calculate the change in nearside fatalities relative to the control group. These "non-occupant fatalities" are the pedestrians, bicyclists and other non-motorists struck and fatally injured by cars of the make-models and model-year ranges defined in Tables 2-1 – 2-4. For any given group of vehicles, the analysis considers the ratio of fatalities who were nearside occupants in the vehicles to the fatalities among pedestrians, bicyclists and other non-motorists who were struck by vehicles from this group. The analysis is based entirely on FARS data.

Two past evaluations of measures to enhance side impact protection – side door beams introduced in 1969-1973 and voluntary TTI(d) improvements of the mid-1980s – also used control groups and FARS data, but the control group was occupant fatalities in frontal impacts.[81] Dalmotas, however, urged NHTSA to substitute non-occupant fatalities for frontals in this

---

[81] Kahane (2004), pp. 138 and 143.

analysis. The two earlier measures were introduced over time periods when frontal protection remained largely unchanged. But the dynamic test for FMVSS 214 was phased in during 1994-1997. Frontal air bags were introduced in some cars at that time or just earlier. They were redesigned in many cars shortly thereafter (1998-1999). Belt use increased. The analysis would be complicated because it would somehow have to adjust the frontal fatalities, perhaps as in Tables 2-8 and 2-9, to account for the effects of those innovations (imprecisely at best).

By contrast, there were no obvious changes to passenger cars during the 1990s that affected their threat to non-occupants. With the non-occupant control group, we may analyze the actual, unadjusted fatality counts from FARS. GES, with its cluster sample design is also not involved. We can use conventional statistical tests, such as chi-square.

A potential problem with this control group, however, is the steady long-term decline of pedestrian and bicyclist fatalities relative to occupants.[82] Basically, people walk less and drive more – for example, as they move away from central cities or further out in the suburbs. In the analysis, the post-standard cars, being newer, will have a somewhat larger share of their crashes in the more recent calendar years, where there are fewer pedestrians. That could elevate the ratio of occupant to non-occupant fatalities for the post-standard cars and depress the effectiveness estimate. Another problem with any control group is a loss of statistical power, because statistics based on the control group contribute to sampling error.

Table 2-12 compares nearside fatalities in multivehicle crashes to non-occupant fatalities, before and after TTI(d) reduction and/or FMVSS 214 certification. As in Section 2.3, the analysis is based on calendar year 1993-2005 FARS data and cars of make-models and model-year ranges in Tables 2-1b, 2-2, 2-3 and 2-4. Thus, the counts of nearside fatalities in multivehicle crashes in Table 2-12 are identical to those in Tables 2-5, 2-6 and 2-7. Non-occupants include PER_TYP = 3, 4, 5, 6, 7, 8 or 19, comprising pedestrians; bicyclists; people in wheelchairs or on roller skates; horseback riders; and occupants of other non-motorized conveyances, animal-drawn conveyances, parked vehicles or work equipment. In most crashes involving non-occupants, there is only one vehicle, and we can include that vehicle in the analysis if it is one of the cars in Tables 2-1b – 2-4. But when there are two or more vehicles, FARS must explicitly indicate through the variable N_MOT_NO that a car in Tables 2-1b – 2-4 was the one that struck the non-occupant.

In the Category 1 make-models with substantial TTI(d) improvements, Table 2-12 shows 1,452 nearside and 1,275 non-occupant fatalities before the improvement, a risk ratio of 1.139. After the improvement, there were 1,096 nearside and 1,057 non-occupant fatalities, a risk ratio of 1.037. That is a 9 percent reduction of nearside relative to non-occupant fatalities. It falls just short of statistical significance at the one-sided .05 level, as evidenced by a chi-square of 2.64 for the 2x2 table created by the first two rows of Table 2-12 (chi-square = 2.70 is needed for significance at that level).

---

[82] *Traffic Safety Facts 2004*, NHTSA Report No. DOT HS 809 919, Washington, 2005, p. 18.

TABLE 2-12

NEARSIDE FATALITIES IN MULTIVEHICLE CRASHES
VS. NON-OCCUPANT FATALITIES
(1993-2005 FARS)

	Nearside Fatalities	Non-Occupant Fatalities	Risk Ratio	Nearside Reduction
Category 1: make-models <u>with</u> substantial TTI(d) improvements				
Before TTI(d) improvement	1,452	1,275	1.139	
After TTI(d) improvement	1,096	1,057	1.037	9 %
Category 2: make-models <u>without</u> TTI(d) improvements				
Before 214 certification	1,486	1,278	1.163	
After 214 certification	1,312	1,044	1.257	− 8 %
Category 3: perhaps major new structure upon 214 certification, TTI(d) change unknown				
Before 214 certification	278	332	.837	
After 214 certification	303	297	1.020	− 22 %
Category 4: no evidence of major new structure upon 214 certification, TTI(d) change unknown				
Before 214 certification	580	612	.948	
After 214 certification	414	429	.965	− 2 %

Nevertheless, the result for Category 1 is more favorable than the other categories; in fact, it is the only positive result in Table 2-12. In Category 2, make-models with unchanged TTI(d), nearside fatalities increased by a non-significant 8 percent relative to non-occupant fatalities after FMVSS 214 certification. In Categories 3 and 4, with smaller N's of cases, the increases were 22 percent (significant at the one-sided .05 level) and 2 percent, respectively.

As discussed above, the long-term decline in non-occupant fatalities could create a "gradient" that increases the ratio of nearside to non-occupant fatalities for the newer, later-model cars. The numbers in Table 2-12 support that hypothesis. Only in Category 1 is the effect of TTI(d) reductions sufficient to overcome the gradient. Two methods are available to adjust for the gradient in the Category 1 results.

One method is to compute the change in risk ratio in Category 1, where TTI(d) was substantially reduced, relative to the corresponding change in Category 2, where we know TTI(d) stayed the same. It is a three-dimensional contingency table analysis in which Category 2 is a "control group on the control group." The risk reduction in Category 1, relative to Category 2 is:

$$1 - \{ [(1096/1057) / (1452/1275)] / [(1312/1044) / (1486/1278)] \} = 16 \text{ percent}$$

This 16 percent fatality reduction is statistically significant at the one-sided .05 level, in fact even at the two-sided .05 level, as evidenced by chi-square = 4.52 for the three-way interaction term when the CATMOD procedure of SAS is applied to the three-dimensional contingency table.[83] It is also fairly close to the 21 percent reduction seen in the basic FARS/GES analyses (Table 2-5),

The other method uses only the data within Category 1 and attempts to adjust, by logistic regression, for various factors that could influence the ratio of nearside to non-occupant fatalities. The data points in the regression are the FARS fatality cases. The dependent variable, NEARSIDE equals 1 for nearside fatalities, 2 for non-occupants. The key independent variable, TTI_IMPR is the TTI(d) status: 0 before the improvement vs. 1 after the improvement. The other independent variables are:

- Above all, the calendar year (CY) of the crash. It will help adjust for the long-term decline in non-occupant fatalities. CY can be entered as a linear variable or alternatively as a categorical variable.

- The age and gender of the driver of the car that was occupied by the nearside fatality or that struck the non-occupant.

- VEHAGE, SPDLIM55, NITE, previously defined in the FARS/GES regressions of Section 2.3.

- Additional variables that may be defined within FARS and tend to interact with the prevalence of non-occupant fatalities:

    o RURAL (= 1 if rural, 0 if urban)

    o FREEWAY (= 1 if limited access divided highway, 0 otherwise)

    o WETROAD (= 1 if surface condition was wet, 0 otherwise)

    o HIFAT_ST (= 1 for the 26 States that have had the highest fatality rates per registered vehicle; 0 for the others)[84]

When CY is entered as a linear variable, the coefficient for TTI_IMPR is -.132. In other words, TTI(d) improvement is associated with a $1 - \exp(-.132) = 12$ percent reduction in nearside relative to non-occupant fatality risk. The coefficient falls just short of statistical significance at

---

[83] For a CATMOD analysis of crash data with statistically significant three-way terms see Morgan, C., *The Effectiveness of Retroreflective Tape on Heavy Trailers*, NHTSA Technical Report No. DOT HS 809 222, Washington, 2001, pp. 29-37, summarized in Kahane (2004), pp. 43-44; *SAS/STAT® User's Guide*, Vol. 1, Version 6, 4th Ed., SAS Institute, Cary, NC, 1990.
[84] Kahane, C.J., *Vehicle Weight, Fatality Risk and Crash Compatibility of Model Year 1991-99 Passenger Cars and Light Trucks*, NHTSA Technical Report No. DOT HS 809 662, Washington, 2003, pp. 24-26.

the one-sided .05 level, as evidenced by a chi-square of 2.56. When CY is instead entered as a categorical variable, results are nearly identical: coefficient = -.131, chi-square = 2.51.

Thus, all of the analyses of nearside fatalities in multivehicle crashes relative to non-occupant fatalities yielded positive point estimates for the effect for TTI(d) improvement in the Category 1 cars. The two effectiveness estimates that attempted to control for the long-term decline in non-occupant fatalities were 16 percent and 12 percent, and the first one is statistically significant.

2.5   Analysis of compact pickup trucks

The dynamic test requirement of FMVSS 214 was extended to LTVs up to 6,000 pounds Gross Vehicle Weight Rating (GVWR), effective on September 1, 1998. TTI(d) may not exceed 85 (same as 4-door cars). In the early 1990s, NHTSA tested 20 production LTVs in side impacts by a MDB. However, only five of the tests were at the future FMVSS 214 speed and conditions (the others were more severe), and only one of those five tested a make-model that would still exist after FMVSS 214 went into effect. That vehicle was a 1991 Toyota compact pickup, and the TTI(d) was 55, substantially better than the 85 the standard would eventually allow. It is the only make-model where the agency can quantitatively compare "before" and "after" performance. Nor did the manufacturers' responses to Information Requests (IR) concerning ten LTVs of model year 1999 provide clear evidence of substantial modifications to meet FMVSS 214.[85]

NHTSA doubts that full-sized pickup trucks, SUVs or vans substantially changed in response to FMVSS 214. Their relatively high, rigid floors, plus the side door beams already in the vehicles must have been were adequate for compliance with the dynamic test. On the other hand, it is conceivable that side structures of compact pickup trucks – smaller, lighter and lower than the full-sized LTVs – were upgraded or padded near the effective date of the dynamic test requirement or somewhat earlier. In his peer review, John Jacobus recommended this report include a statistical analysis of compact pickup trucks.

Five manufacturers produced compact pickup trucks for sale in the United States before as well as after September 1, 1998:

- Toyota is the only make-model tested before and after FMVSS 214. TTI(d) was 55 before, far better than the FMVSS 214 requirement, but it did not improve afterwards.

- Dodge Dakota, Ford (sold as Ford Ranger or Mazda B) and Nissan pickup trucks received integrated platform redesigns a year or two before September 1, 1998: Dodge Dakota in model year 1997 and the others in model year 1998. Any FMVSS 214-related modifications were presumably "built in" during the redesign, for there is little evidence of subsequent change for model year 1999.

- GM's compact pickup trucks in those years were sold by Chevrolet (S/T), GMC (Sonoma) and Isuzu (Hombre). There was no integrated platform redesign in the mid-to-

---

[85] *Federal Register* 60 (July 28, 1995): 38749; *Preliminary Economic Assessment, NPRM for Light Trucks, Buses and Multipurpose Passenger Vehicle, Dynamic Side Impact Protection, FMVSS No. 214*, NHTSA Docket No. 88-06-N23-001, 1994, Chapter III.

late 1990s. Jacobus found a GM advertisement in *Motor Trend*, July 1998, stating that all the 1999 Chevrolet and GMC cabs in compact pickup trucks were improved with: (1) stiffened cab joints, (2) deeper rocker sections and (3) the addition of a magnesium beam within the instrument panel that ties the A-Pillars together, increasing lateral stiffness. These are evidently FMVSS 214-type upgrades. It is not clear, without additional detail, if they should be called "major" or "minor" structural upgrades.

In other words, Toyota's pickup trucks are like the Category 2 passenger cars: TTI(d) was known before and after FMVSS 214, and it did not improve. They are excluded from the analysis. The others are like Category 4. They were tested after FMVSS 214; TTI(d) scores, adjusted to the standard's 33.54 impact speed, were all substantially better than 85. But TTI(d) before FMVSS 214 is unknown, and it is not clear how extensively the vehicles were modified. As with passenger cars, the analysis compares the last three "before" model years to the first three "after." Based on the above discussion on when vehicles may have been modified, the following model years are included in the statistical analysis:

	"Before"	"After"
Dodge Dakota	1994-1996	1997-1999
Ford Ranger and Mazda B	1995-1997	1998-2000
Chevrolet S/T, GMC Sonoma and Isuzu Hombre	1996-1998	1999-2001
Nissan pickup and Frontier	1995-1997	1998-2000

As in Table 2-12, nearside fatalities in multivehicle crashes are analyzed relative to a control group of non-occupant fatalities in 1993-2005 FARS. Table 2-13 does not show a statistically meaningful change in nearside fatality risk after FMVSS 214. The risk ratio increased by a non-significant 5 percent from .431 to .451. The available N of nearside fatalities is substantially smaller than in the corresponding analyses of passenger cars in Table 2-12, precluding more detailed analyses that would subdivide the population of compact pickup trucks (e.g., 2-wheel-drive vs. 4-wheel-drive, conventional cab vs. extended cab).

TABLE 2-13

COMPACT PICKUP TRUCKS: NEARSIDE FATALITIES
IN MULTIVEHICLE CRASHES VS. NON-OCCUPANT FATALITIES
(1993-2005 FARS)

	Nearside Fatalities	Non-Occupant Fatalities	Risk Ratio	Nearside Reduction
Before TTI(d) improved	256	594	.431	
After TTI(d) perhaps improved	162	359	.451	– 5 %

Even when N is enlarged by extending the model-year range to ± 4 years or ± 5 years, the point estimates remain nearly the same (5 and 7 percent increases, respectively) and they are not statistically significant.

It is noteworthy that the risk ratios of nearside to non-occupant fatalities in Table 2-13 (.431, .451) are less than half the corresponding ratios for passenger cars (ranging from .837 to 1.257 in Table 2-12). In part, that is because compact pickup trucks are intrinsically less vulnerable than cars in side impacts. But occupants of pickup trucks also tend to be younger and more often male – and, as a result, at lower risk than car occupants.

In short, lack of information about pre-standard performance and insufficient nearside fatality cases ruled out any conclusions about the effect of FMVSS 214 in compact pickup trucks.

2.6 Best effectiveness estimates for passenger cars

All of the analyses for the 15 Category 1 make-models of passenger cars showed lower fatality risk for nearside occupants in multivehicle crashes after the substantial TTI(d) improvement.

- The most direct and basic analysis is the first one in this chapter, the comparison of fatality rates per 1,000 towaway-involved occupants, based on FARS and GES data, in Table 2-5. It estimated a statistically significant 21 percent fatality reduction after the TTI(d) improvement.

- Another important analysis in Section 2.3 is the regression of fatality reduction by TTI(d) improvement, with the data aggregated at the make-model level. It showed that the greater the TTI(d) improvement, the greater the fatality reduction; the estimated effect was equivalent to a statistically significant 17 percent fatality reduction for the 23-unit TTI(d) improvement in Category 1.

- The analyses of nearside vs. non-occupant fatalities in Section 2.4 have the advantage of using only FARS data. The three-dimensional contingency table analysis, essentially controlling for calendar year, showed a statistically significant 16 percent nearside fatality reduction for the Category 1 models.

The average of those three key results, **18 percent**, will serve as the best point estimate of the fatality reduction in Category 1.[86] Confidence bounds for that estimate should take into account not only sampling error (the result of limited data) but computational uncertainty (as evidenced by varied results when different analyses are applied to fundamentally the same data). The 90 percent confidence bounds range from **7 to 28 percent**.[87]

---

[86] The actual effectiveness estimates (not rounded) are 21.30%, 16.87% and 15.76%.

$1 - \exp \{ [\log(1 - .2130) + \log(1 - .1687) + \log(1 - .1576)] / 3 \} = 1 - \exp(-.1986) = 18.01$ percent

[87] When expressed as log r, the three estimates are -.2395, -.1848 and -.1715 and have standard errors .0521, .0889 and .0807, respectively. The confidence bounds will use the lowest and highest point estimates (computational error) and the tightest standard error (sampling error). The lower bound is $1 - \exp(-.1715 + 1.833 \times .0521) = 7\%$. The upper bound is $1 - \exp(-.2395 - 1.833 \times .0521) = 28\%$. (−1.833 is the 5th percentile of a t-distribution with 9 degrees of freedom, the appropriate multiplier for the sampling error of the first estimate.) This approach tries to address Dalmotas' guidelines for the error analysis.

The preceding estimate pertains to nearside occupants in multivehicle crashes. But how about other people exposed to side impacts, specifically nearside occupants in single-vehicle crashes, and farside occupants? Intuitively, the structure and padding used to improve TTI(d) could have some influence in nearside single-vehicle crashes. Many of these crashes, however, are impacts into poles or trees and are quite different from the scenario in the FMVSS 214 test; added structure might do little in those cases. It is difficult to envision much effect in farside impacts; there might be some improvement in structural integrity of cars that received extensive reinforcement of cross-members. Intuitive expectations should be compared to statistical results: evaluations of earlier side-impact protection showed some benefits where they were not necessarily expected (e.g., in farside crashes for side door beams; in single-vehicle crashes for the first round of voluntary TTI(d) improvements in 2-door cars).

This time, the statistical results are more consistent with intuitive expectations. For all types of side impacts **other than** nearside impacts by another vehicle, the fatality rate was 7 percent lower in the Category 1 cars after the TTI(d) improvement, based on a FARS-GES data similar to Table 2-5. The reduction is not statistically significant (t = -1.03 in the analysis of sampling error similar to the procedure for Table 2-5). It is much lower than fatality reduction in nearside multivehicle crashes. In just the nearside single-vehicle impacts, the reduction was merely 3 percent. While it is conceivable that TTI(d) improvement could have some benefits in side impacts other than nearside multivehicle, there is not enough evidence to uphold a specific, positive effectiveness.

The 18 percent "best estimate" of fatality reduction (18.01 percent, not rounded) was estimated specifically for 15 make-models known to have substantially improved TTI(d), the Category 1 cars listed in Tables 2-1a and 2-1b. How does that scale down to the overall car fleet, which includes some make-models with little or no change in TTI(d) over time?

Tables 2-1a and 2-1b indicate that the registration-weighted average TTI(d) improvement for those 15 make-models was 23 units. Because

$$\log r = \log (1 - .1801) = -.1986$$

log r is reduced by .1986 / 23 = .00863 per unit, corresponding to a **0.863 percent fatality reduction per unit reduction of TTI(d)** (confidence bounds, 0.33 to 1.46 percent fatality reduction per unit[88]).

The concept of "fatality reduction per unit reduction of TTI(d)" should be introduced with some words of caution.[89] This chapter enumerates 15 make-models that substantially upgraded and/or padded side structures (as described explicitly in response to Information Requests or inferred from a large TTI(d) improvement). The statistical analyses show these same 15 make-models, as a group, experienced a statistically significant 18 percent fatality reduction for nearside occupants in multivehicle crashes. They also experienced a 23-unit reduction in average

---

[88] The lower bound is (-.1715 + 1.833x.0521) / 23 = -.00330. The upper bound is (-.2395 - 1.833x.0521) / 23 = -.01457.
[89] Emphasized by Dalmotas in his peer review.

TTI(d).[90] The effect per unit reduction is computed for the purpose of obtaining a scaling factor: if another group of cars – the entire fleet, for example – uses **qualitatively similar methods** to upgrade side structures as these 15 models, the fatality reduction[91] for that group will be greater or smaller than 18 percent in the proportion that its average TTI(d) improvement is greater or smaller than 23 units. As Dalmotas points out, this is not as strong as claiming that **any method** improving TTI(d) by one unit will necessarily reduce fatality risk by 0.863 percent; such a claim seems beyond what could be inferred from statistical analyses of crash data for production cars, because the analysis focuses on the upgrade methods actually implemented in those cars.

The result is quite close to the 0.927 percent fatality reduction per unit improvement of TTI(d) (confidence bounds, 0.52 to 1.33 percent) found in the Phase 1 evaluation report for the first round of voluntary TTI(d) improvements in pre-standard, 2-door cars of model years 1981-1993.[92]

- The 0.927 estimate was calibrated from crash data for a cross-section of 2-door make-models of model years 1981-1993; 16 of the 17 make-models in that analysis (94 percent) had TTI(d) over 90; none were FMVSS 214-certified.

- The 0.863 estimate is based on "before and after" crash data for 15 make-models that substantially improved TTI(d). These models' TTI(d) averaged 85 before and 62 after the improvement. The model years of the cars in the analysis ranged from 1991 to 2002; 80 percent of the cars had TTI(d) less than 90; 70 percent were 4-door cars; and 58 percent of the cars were FMVSS 214-certified.

When calculating the fleet-wide, long-term effect of side impact protection, we will assume a 0.927 percent effect per unit of TTI(d) improvement when TTI(d) is above 90, and the 0.863 percent effect per unit improvement when TTI(d) is 90 or less. More realistically, of course, the effect would not change abruptly at 90 or any specific other point but would transition smoothly in some range well above and below 90; however, we have no basis for establishing a transition rate or a range. Since we are selecting a specific "boundary," 90 is an intuitively good choice, even though the datasets in the two analyses slightly overlap on both sides of 90, because it is the FMVSS 214 requirement for 2-door cars. Essentially, most 2-door cars until just before FMVSS 214 had TTI(d) over 90, whereas most 4-door cars even before the standard, and of course all 214-certified cars had scores below 90.

In **four-door cars**, Table 1-4 indicated that the average TTI(d) of baseline, model year 1981-1985 cars was 85, and the average for cars certified to FMVSS 214, but not equipped with side air bags was 63: an average improvement of 22 units for the entire fleet of 4-door cars. The entire improvement is in the below-90 range. Given that fatality risk is reduced by 0.863 percent per unit improvement of TTI(d), a 22-unit improvement corresponds to an overall **17 percent**

---

[90] Moreover, the regression of fatality reduction by TTI(d) improvement, with data aggregated by make-model, shows a significant correlation between the amounts of TTI(d) improvement and fatality reduction, producing some confidence that the observed fatality reduction is not an artifact of the specific make-models included.
[91] When expressed as log r.
[92] Kahane (1999), p. 84. The regression coefficient 0.927 had standard error .248; 90% confidence bounds are .927 ± 1.645 x .248. Note, however, that the 0.927 estimate in the Phase 1 report applies to all side impacts, whereas the 0.863 effect in this report is only claimed for nearside impacts by another vehicle.

fatality reduction[93] in nearside impacts by another vehicle (confidence bounds, 7 to 27 percent), for 214-certified cars relative to 1981-1985 baseline cars.

For **two-door cars**, Table 1-4 indicated that the average TTI(d) of baseline, model year 1981-1985 cars was 114, and the average for cars certified to FMVSS 214, but not equipped with side air bags was 69: an average improvement of 45 units for the entire fleet of 2-door cars. However, the first 24 units of this improvement were in the 90+ range, and the last 21 units were in the below-90 range. Fatality risk is reduced by 0.927 percent per unit improvement from 114 to 90, and by 0.863 percent per unit improvement from 90 to 69. The overall fatality reduction since 1981-1985 in multivehicle, nearside impacts has been **33 percent**[94] (confidence bounds, 18 to 47 percent), with smaller reductions in single-vehicle nearside impacts and in farside impacts.

---

[93] $1 - \exp(-.00863 \text{x} 22)$.

[94] $1 - [(1 - .00927)^{24} \text{ x } \exp(-.00863 \text{x} 21)]$.

CHAPTER 3

EFFECT OF SIDE AIR BAGS
ON FATALITIES AND EJECTION IN SIDE IMPACTS

3.0   Summary

Side air bags designed for torso protection began to appear in 1996 passenger cars. Nearly 30 percent of model year 2001-2003 cars were equipped with them. Head-protection air bags first appeared on 1998 cars and, by 2003, were installed on nearly 20 percent of new cars. The current evidence is that the combination of torso bags and head protection reduces fatality risk in nearside impacts by a statistically significant 24 percent for drivers and right-front passengers. It is less clear how much of this combined effect is due to the torso bag, and how much to head protection. Most of the analyses suggest that torso and head air bags are both effective and perhaps contribute more or less equally to the combined effect.

3.1   Car models that received standard or optional side air bags

The first steps of the analysis are to identify the cars with side air bags in the front seats, especially those make-models that shifted from no air bags to having some type of side air bags as standard equipment, or in substantial quantities as optional equipment. These are the make-models that allow a direct comparison of fatality risk with and without the air bags, either on a "before-after" basis (if standard equipment) or on a cross-sectional basis (if optional equipment).

The Insurance Institute for Highway Safety (IIHS) has published several reports with accurate information about what cars and LTVs have side air bags, and what types of bags.[95] Other information sources include NHTSA's own brochures on *Buying a Safer Car*, lists of standard and optional equipment by make-model, sub-series and model year at www.cars.com, and the Passenger Vehicle Identification Manuals of the National Insurance Crime Bureau (NICB).[96] These sources were used to compile a SAS program that identifies if a vehicle has side air bags, and what type, based on:

- The make-model and model year, if the same type of side air bags (or none at all) was standard equipment on all vehicles of that make-model and model year.

- The Vehicle Identification Number (VIN), if side air bags were installed on some but not all vehicles of a given make-model and model year, or different types were installed.

---

[95] *Status Report*, Vol. 36, January 6, 2001, Vol. 38, June 28, 2003 and August 26, 2003, Insurance Institute for Highway Safety, Arlington, VA; Braver, E.R. and Kyrychenko, S.Y., *Efficacy of Side Airbags in Reducing Driver Deaths in Driver-Side Collisions*, Insurance Institute for Highway Safety, Arlington, VA, 2003.

[96] *Buying a Safer Car 2000*, NHTSA Publication No. DOT HS 809 046, Washington, 2000; *Buying a Safer Car 2001*, NHTSA Publication No. DOT HS 809 152, Washington, 2000; *Buying a Safer Car 2002*, NHTSA Publication No. DOT HS 809 409, Washington, 2002; *Buying a Safer Car*, NHTSA Publication No. DOT HS 809 546, Annual publication, 2003-2005; *Passenger Vehicle Identification Manual*, Annual Publication, National Insurance Crime Bureau, Palos Hills, IL.

The program also notes if a vehicle might have had optional air bags, but it cannot be determined from the VIN.

The types of side air bags are:

- Torso bag only.

- Combination torso/head air bag that deploys from the seat to protect the torso but also extends upward far enough to protect the head impact zone around the side window.

- Torso bag plus separate head curtain or inflatable tubular structure that deploys from the roof rail.

- Head curtain only, without torso bag.

Table 3-1 identifies the "core" group of make-models with some type of side air bags as standard equipment in certain model years up to 2003, and that originally did not have any side air bags or, at least, shifted from one type of side air bags as standard equipment in one year to another type in another year (but make-models that always had torso bags and head protection, shifting only from combination bags to torso bags plus separate head curtains, are not included).[97] All of the cars in Table 3-1 certified to the dynamic test requirements of FMVSS 214 and were equipped with dual frontal air bags and manual 3-point belts in the front seats. For example, Lincoln Town Car was initially certified to FMVSS 214 in model year 1994 and did not have any side air bags in 1994-1998. In 1999-2003, all were equipped with combination torso/head air bags at the driver and right-front (RF) passenger seats. Buick LeSabre shifted from no side air bags (1997-1999) to torso bags only (2000-2002). Volkswagen Passat shifted twice, from no side air bags in 1995-1997 to torso bags only in 1998-2000 to torso bags plus head curtains in 2001-2003. Volvo V70, a more recently introduced make-model, started out with torso bags in 1998 and added head protection in 1999. Models receiving head curtains usually went through an interim period with torso bags only, while those receiving combination bags often did not.

These core make-models are especially suitable for statistical analyses because all sales in one model year have one type of side impact protection, and all sales in the next year have a different type. There is no selection bias of more safety-conscious owners choosing optional air bags. Another advantage is the possibility of using data from the General Estimates System (GES) of the National Automotive Sampling System (NASS), where the make-model is known for 86 percent of the cases, even though the VIN itself is only known for 73 percent. The right column of Table 3-1 shows the years when make-models were substantially redesigned. In the majority of cases, the transition to side air bags, or from one type of air bags to another did not coincide with major redesigns: the cars remained largely the same, except for the change in the side air bags. Table 3-1 includes 40-50 models with side air bags as standard equipment at one time or another.[98] The data are not "clustered" into a small number of make-models.

---

[97] A few make-models that had only one type of side air bags, or that only shifted from combination bags to torso bags plus head curtains are included in Table 3-1 because they are fairly similar to other make-models produced by the same manufacturer: Audi S4 and S6 (similar to A4 and A6); Volvo 40-series, 60-series and 80-series (can be considered replacements, with new names, of earlier Volvo make-models).

[98] The range depends on whether or not you separately count the individual models listed under Audi, Porsche and Volvo.

TABLE 3-1: MAKE-MODELS WITH STANDARD SIDE AIR BAGS IN THE FRONT SEATS

(**All cars listed here are 214-certified and equipped with dual frontal air bags**; make-models that had the same type of side impact protection in every year are listed separately in Table 3-4 and not included in the statistical analyses)

Make-Model	None	Torso Only	Torso/Head Combination Bags	Torso + Head Curtains	Head Curtains Only	MY with Major Redesign
Lincoln Town Car	1994-1998		1999-2003			none
Lincoln Continental	1995-1998		1999-2002			none
Buick LeSabre[99]	1997-1999	2000-2002				2000
Buick Park Avenue	1997-1999	2000-2003				none
Cadillac DeVille	1994-1996	1997-2003				2000
Cadillac Seville						1998
Drivers	1994-1997	1998-2000	2001-2003			
RF passengers	1994-1997	1998-2003				
Oldsmobile Aurora	1997-1999	2001-2003				2001
Pontiac Bonneville	1997-1999	2000-2003				2000
Saturn LS/LW[100]	*2000*				*2002-2003*	*none*
Volkswagen Jetta[101]	1995-1998	2000		2001-2003		mid 99
Volkswagen Golf[102]	1995-1998	2000		2001-2003		mid 99

---

[99] Torso bags optional in 2003.
[100] Head-curtain-only optional in 2001. Saturn L is not included in the analyses because it is the only make-model with standard head-curtain-only, and there are not enough cases for statistically meaningful analyses.
[101] In 1999, torso air bags were standard on the New Jetta only. Also, in 1998-1999, a small percentage of the earlier Jetta had optional torso air bags.
[102] In 1999, torso air bags were standard on the New Golf only. Also, in 1998-1999, a small percentage of the earlier Golf had optional torso air bags.

TABLE 3-1 (Continued): MAKE-MODELS WITH STANDARD SIDE AIR BAGS IN THE FRONT SEATS

(All cars listed here are 214-certified and equipped with dual frontal air bags; make-models that had the same type of side impact protection in every year are listed separately in Table 3-4 and not included in the statistical analyses)

Make-Model	None	Torso Only	Torso/Head Combination Bags	Torso + Head Curtains	Head Curtains Only	MY with Major Redesign
Volkswagen Cabrio[103]	1995-1998	2000-2002				none
Volkswagen Passat	1995-1997	1998-2000		2001-2003		1998
Audi A4, S4, A6, S6, A8						
A4[104]	1996-1997	1998-1999		2001-2003		2002
S4				2000-2002		none
A6[105]	1995-1997	1998-1999		2001-2003		1998
S6				2002-2003		none
A8	1997-1999	1997-1999		2000-2003		none
BMW 300[106]						
4-door	1995-1997	1998		1999-2003		1999
2-door coupe/HB	1995-1997	1998-1999		2000-2003		2000
BMW 500[107]	1995-1996	1997		1998-2003		none
BMW 700[108]	1995-1996	1997		1998-2003		none
BMW Z3	1996-1997	1999-2002				none
Jaguar XK coupe	1997-2000		2001-2003			none

---

[103] Torso air bags optional in 1999.
[104] Mix of torso-only and torso + head curtains in 2000.
[105] Mix of torso-only and torso + head curtains in 2000.
[106] BMW are equipped with inflatable tubular structures rather than head curtains.
[107] BMW are equipped with inflatable tubular structures rather than head curtains.
[108] BMW are equipped with inflatable tubular structures rather than head curtains.

TABLE 3-1 (Continued): MAKE-MODELS WITH STANDARD SIDE AIR BAGS IN THE FRONT SEATS

(All cars listed here are 214-certified and equipped with dual frontal air bags; make-models that had the same type of side impact protection in every year are listed separately in Table 3-4 and not included in the statistical analyses)

Make-Model	None	Torso Only	Torso/Head Combination Bags	Torso + Head Curtains	Head Curtains Only	MY with Major Redesign
Jaguar XJ Sedan	1996-1997	1998-2003				none
Mazda Millenia	1995-2000		2001-2002			none
Mercedes C Sedan	1995-1997	1998-2000		2001-2003		2001
Mercedes CL Coupe	1998-1999	2000		2001-2003		2000
Mercedes E Sedan	1995-1996	1997-1998		1999-2003		1996,2003
Mercedes S Sedan	1995-1996	1997-1999		2000-2003		2000
Mercedes SL	1994-1996	1997-2002	2003			2003
Porsche						
911	1995-1998	1999,02-03	2000-2001			1999
Boxster	1997	98-99,02-03	2000-2001			none
Toyota Avalon	1995-1997	1998-2003				none
Volvo						
850,940,960	1994-1995	1996-1997				none
40-series			2000	2001-2003		none
60-series				2001-2003		none
C 70		1998-1999	2000-2003			none
S 70		1998	1999-2002			none
V 70		1998	1999-2000	2001-2003		2001
80-series				1999-2003		none
90-series		1998				none

TABLE 3-1 (Concluded): MAKE-MODELS WITH STANDARD SIDE AIR BAGS IN THE FRONT SEATS
(**All cars listed here are 214-certified and equipped with dual frontal air bags**; make-models that had the same type of side impact protection in every year are listed separately in Table 3-4 and not included in the statistical analyses)

Make-Model	None	Torso Only	Torso/Head Combination Bags	Torso + Head Curtains	Head Curtains Only	MY with Major Redesign
Acura TL	1996-1999	2000-2003				1999
Acura RL	1997-1998	1999-2003				none
Acura CL	1997-1999	2001-2003				none
Hyundai Sonata	1995-1998	1999-2000	2001-2003			none
Hyundai Elantra	1997-2000		2001-2003			2001
Hyundai Tiburon	1997-2001		2003			2003
Infiniti Q45[109]		1998	1999-2001	2002-2003		2002
Infiniti I30/I35	1996-1997	1998-1999	2000-2003			2002
Lexus ES	1994-1997	1998-2001		2002-2003		1997,2002
Lexus LS	1995-1996	1997-2000		2001-2003		2001
Lexus SC	1995-2001	2002-2003				2001
Lexus GS	1996-1997	1998-2000		2001-2003		1998
Lexus IS	2001	2001		2002-2003		none

---

[109] Torso air bags optional in 1997.

74

The core group of cars can be supplemented with additional make-models that, in certain years, offered a choice between side air bags and no side air bags or a choice between two different types of side air bags. The "choice" may be an option entirely up to the buyer, unrelated to other equipment on the car, or it may be standard equipment on certain sub-series and unavailable on others. As stated above, analyses including these models could be affected by selection bias, such as more safety-conscious owners choosing optional air bags.

In either case, the type of side impact protection must be decodable from the VIN. Also, for comparison purposes, there ought to be substantial numbers of cars with each alternative type of protection: let us say side air bags must be present, in each make-model, in at least 1/6 and at most 2/3 of the cars. Table 3-2 lists 17 new make-models, not cited in Table 3-1, meeting those criteria. For example, Chrysler Concorde, 300M and PT Cruiser were available with combination bags protecting the head and torso or without side air bags in 2001-2003. We may also include additional model years for models in Table 3-1 when side air bags were optional, as indicated in the footnotes for Table 3-1 (e.g., Buick LeSabre in 2003). All of the cars are FMVSS 214-certified and equipped with dual frontal air bags. For the models in Table 3-2, only the range of model years listed in Table 3-2 is included in any analysis.[110] The supplementary make-models in Table 3-2 (and the supplementary years for the models in Table 3-1) are only included in analyses based entirely on the Fatality Analysis Reporting System (FARS), where the VIN is retrieved by the crash investigators from State records and recorded on the file for 98 percent of the vehicle cases. They are not included in analyses that use GES data, where VIN is recorded only in 73 percent of the cases.

As in Chapter 2, we may sometimes want a "control group" of make-models that did not change in terms of side impact protection. Table 3-3 lists 19 make-models that never had side air bags, or that did not have any before 2003. The model year ranges included in Table 3-3 – e.g., 1996-2003 for Ford Mustang – start when the make-model was first certified to FMVSS 214.

Table 3-4 enumerates other make-models that were partially or fully equipped with side air bags, but will not be included in the analyses of this chapter. Part A of Table 3-4 lists make-models that had only a small proportion of optional side air bags, well below 15 percent. For these models, statistically meaningful fatality rates with air bags cannot be computed, and if these models were mixed with the others (Tables 3-1 and 3-2), they would skew the sample. Part B specifies make-models such as Honda Civic or Mitsubishi Galant, where side air bags are optional and their presence or absence on individual vehicles can never or hardly ever be decoded from the VIN. Part C enumerates recently introduced make-models that were equipped with side air bags as standard equipment from the start, and furthermore did not switch from torso-only to torso bags plus any type of head protection. Here, there is no comparison group of any kind.

---

[110] Exception: Honda Accord. In 2000-2003, some sub-series have side air bags as standard equipment, but in most other sub-series, the presence or absence of side air bags cannot be established from the VIN. Only a small number of cars can be identified that definitely did not have side air bags. Therefore, we have included 1998-1999, when none of the cars were equipped with side air bags, to enlarge the comparison group without the bags.

TABLE 3-2: MAKE-MODELS WITH SUBSTANTIAL PROPORTIONS OF OPTIONAL[111] SIDE AIR BAGS IN THE FRONT SEATS
(Side impact protection must be decodable from VIN;
side air bags must be present in at least 1/6 and at most 2/3 of the cars;
all cars listed here are 214-certified and equipped with dual frontal air bags)

Make-Model	MY with Optional Side Air Bags	Choices Available
Chrysler Concorde, 300M	2001-2003	Combination, none
Chrysler PT Cruiser	2001-2003	Combination, none
Buick LeSabre	2003	Torso only, none
Buick Century (driver only)	2000-2003	Combination, none
Buick Regal (driver only)	2000-2003	Combination, none
Cadillac Catera[112]	1997-2001	Torso only, none
Chevrolet Impala (driver only)	2000-2003	Combination, none
Chevrolet Monte Carlo (driver only)	2000-2003	Combination, none
Nissan Maxima	1998-1999	Torso only, none
	2000-2003	Combination, none
Nissan Altima	2000-2001	Combination, none
	2002-2003	Torso + Curtain, none
Nissan 350Z	2003	Torso + Curtain, none
Honda Accord[113]	1998-2002	Torso only, none
	2003	Torso + Curtain, Torso only, none
Subaru Legacy	2000-2003	Torso only, none
Toyota Corolla	1998-1999	Torso only, none
Toyota Camry (including Solara)	1998-2001	Torso only, none
Toyota Camry (excluding Solara)	2002-2003	Torso + Curtain, none
Mitsubishi Eclipse	2003	Torso only, none
Infiniti Q45	1997	Torso only, none

---

[111] Includes make-models where side air bags are standard equipment in selected sub-series and unavailable in other sub-series.

[112] Torso bags are standard in 2000-2001 and optional in 1997-1999.

[113] Model years 1998-2003 provide a good mix of VIN-decodable cars with and without side air bags. In 1998-1999, Accord did not have any side air bags; in 2000, a mix of torso-only, none, and non-decodable; in 2001-2002, a mix of torso-only and non-decodable, and in 2003, a mix of torso + curtain, torso-only and none.

TABLE 3-3: CONTROL-GROUP MAKE-MODELS
WITHOUT SIDE AIR BAGS BEFORE 2003
(All 214-certified with dual frontal air bags; at least 6 model years without side air bags)

Make-Model	MY without Side Air Bags
Ford Mustang	1996-2003
Ford Escort	1997-2002
Ford Crown Victoria	1994-2002
Mercury Grand Marquis	1994-2002
Cadillac Eldorado	1994-2002
Chevrolet Corvette	1997-2003
Chevrolet Camaro	1995-2002
Chevrolet Cavalier	1997-2002
Chevrolet Malibu	1997-2003
Olds Achieva/Alero	1997-2003
Pontiac Firebird	1995-2002
Pontiac Sunfire	1997-2002
Pontiac Grand Am	1997-2003
Pontiac Grand Prix	1995-2003
Mazda Miata	1997-2003
Mitsubishi Mirage	1997-2002
Mitsubishi Diamante	1997-2003
Suzuki Esteem	1997-2002
Hyundai Accent	1996-2002

TABLE 3-4: OTHER MAKE-MODELS WITH SIDE AIR BAGS,
NOT INCLUDED IN THE ANALYSES

A. Make-models with small (< 15 percent) proportions of optional side air bags[114]

Make-Model	MY with Optional Side Air Bags	Choices Available
Chrysler LHS	2001	Combination, none
Chrysler Sebring	2001-2003	Head curtain only, none
Dodge Neon	2001-2003	Combination, none
Dodge Intrepid	2001-2003	Combination, none
Dodge Stratus	2001-2003	Head curtain only, none
Ford Crown Victoria	2003	Combination, none
Ford Taurus	2001-2003	Combination, none
Ford Focus	2000-2003	Combination, none
Mercury Grand Marquis	2003	Combination, none
Mercury Sable	2001-2003	Combination, none
Mercury Cougar	1999-2002	Combination, none
Mercury Marauder	2003	Combination, none
Chevrolet Cavalier	2003	Torso only, none
Chevrolet Prizm	1998-2002	Torso only, none
Pontiac Vibe	2003	Torso only, none
Saturn SL/SC	2001-2002	Head curtain only, none
Saturn Ion	2003	Head curtain only, none
Nissan Sentra	2001-2003	Combination, none
Mazda Protégé	2001-2003	Combination, none
Mazda 626	2000-2002	Combination, none
Mazda 6	2003	Torso + Curtain, none
Subaru Impreza[115]	2002-2003	Torso only, none
Subaru Outback	2003	Torso only, none
Toyota Corolla	2000-2003	Torso only, none
Toyota Celica	2000-2003	Torso only, none
Toyota Camry Solara	2002-2003	Torso only, none
Toyota Echo	2001-2003	Torso only, none
Toyota Prius	2001-2003	Torso only, none
Toyota Matrix	2003	Torso only, none

---

[114] Another reason Chrysler LHS, Mercury Marauder, Chevrolet Cavalier, Pontiac Vibe, Saturn Ion, Mazda 6, Subaru Outback, Toyota Echo and Toyota Matrix are excluded is that there are not yet any FARS cases of vehicles equipped with side air bags.

[115] Fewer than 15 percent are without torso air bags.

TABLE 3-4 (Concluded): OTHER MAKE-MODELS WITH SIDE AIR BAGS,
NOT INCLUDED IN THE ANALYSES

B. Make-models with optional side air bags that cannot be identified from the VIN

Make-Model	MY with Optional Side Air Bags	Choices Available
Honda Civic	2001-2003	Torso only, none
Mitsubishi Galant	1999-2003	Torso only, none
Mitsubishi Eclipse	2000-2002	Torso only, none
Mitsubishi Lancer	2002-2003	Torso only, none
Hyundai Accent	2003	Combination, none

C. Make-models that always had standard side air bags, all torso-only or all torso + some type of head protection (there are no comparison vehicles of the same make-model)[116]

Make-Model	MY with Standard Side Air Bags	Type of Side Air Bag
Ford Thunderbird[117]	2002-2003	Combination
Lincoln LS	2000-2003	Combination
Cadillac CTS	2003	Torso + Curtain
Volkswagen Beetle	1998-2003	Torso only
Audi TT	2000-2003	Combination
Audi S8, Allroad	2001-2003	Torso + Curtain
Mini-Cooper	2002-2003	Torso + Inflatable Tubular Structure(?)
BMW Z4, Z8	2000-2003	Torso only
Jaguar S Type	2000-2003	Combination, Torso +Curtain
Jaguar X Type	2002-2003	Torso + Curtain
Mercedes SLK	1998-2003	Torso only
Mercedes CLK	1998-2003	Torso only
Saab 9-3	1999-2003	Combination, Torso +Curtain
Saab 9-5	1999-2003	Combination
Acura RSX	2002-2003	Torso only
Hyundai XG300	2001-2003	Combination
Infiniti G20/G35	1999-2003	Combination, Torso + Curtain
Infiniti M45	2003	Torso + Curtain
Kia Optima	2001-2003	Combination

---

[116] Another reason Audi S8 and BMW Z8 are excluded is that there are not yet any FARS cases.
[117] The Ford Thunderbird without side air bags was discontinued in 1997. It is not a good comparison vehicle because it was discontinued five years before the new one was introduced in a quite different body style.

3.2 Basic analyses of torso and head air bags in nearside impacts of cars

Three combinations of data and analysis methods will be used and will provide at least three parallel effectiveness estimates for the principal analysis questions:

1. Fatality rates per 1,000 nearside, front-seat occupants in police-reported side impacts (i.e., drivers in left-side impacts and right-front passengers in right-side impacts) may be computed for the "core" group of make-models in Table 3-1 and compared for the model years when these cars had no side air bags, standard torso bags only, and standard torso bags plus head protection. As in Section 2.3 (evaluation of TTI(d) improvements), fatality rates are generated from two data files: the numerator – the number of fatalities – from the Fatality Analysis Reporting System (FARS), and the denominator – the number of exposed occupants – from the General Estimates System (GES) of the National Automotive Sampling System (NASS).

2. Fatalities in directly frontal and rear impacts are a control group; the ratio of nearside fatalities to control-group fatalities is computed for the make-models in Table 3-1 and compared for cars without side air bags, with standard torso bags, and with standard torso bags plus head protection. The analysis, based on FARS data alone, resembles the approach in Section 2.4.

3. The analysis method is identical to the preceding one, but the make-models in Table 3-1 are supplemented by the make-models and model years in Table 3-2, which offered optional side air bags (and the VIN identifies whether or not a specific car was equipped with the bags).

FARS-GES analyses

FARS is a census of the nation's fatal crashes, but it excludes crashes where nobody died. GES is a probability sample of the nation's crash involvements, and when GES cases are weighted by the inverse sampling fractions they generate unbiased estimates of national totals. However, the number of fatality cases within GES itself is small. Statistically more accurate fatality rates are obtained by using the FARS census data for the numerator and GES data for the denominator.[118]

The analysis is based on calendar year 1993-2005 FARS and GES data. The make-model of a car can be identified in GES by decoding the VIN and/or from GES' own make-model codes. Approximately 14 percent of GES cases have missing VIN and make-model codes, but since that percentage is the same with and without side air bags (i.e., does not vary with vehicle age or by calendar year) it should not influence the effectiveness estimate.[119] Other analysis variables, such as the VIN on FARS and the impact location on both files have less than 3 percent missing data.

---

[118] For examples of analyses combining FARS and GES see Joksch, H., *Vehicle Design versus Aggressivity*, NHTSA Technical Report No. DOT HS 809 184, Washington, 2000.

[119] The VIN itself, however, is missing in 27 percent of the cases and that makes it not advisable to extend the FARS-GES analyses to the models in Table 3-2, where the presence or absence of side air bags must be decoded from the VIN.

The analyses include single- as well as multivehicle crashes because, intuitively, side air bags could be effective in both types of crashes. In GES data, "nearside occupants" include drivers when the principal impact point is on the left side and right-front passengers when the principal impact is on the right side. Likewise, in FARS data, "nearside occupants" include drivers when the principal impact point is 8, 9 or 10:00 and right-front passengers when the principal impact is 2, 3 or 4:00.[120]

Confounding of "towaway" with side air bags Past analyses of crash data have often computed fatality or injury rates per 1,000 occupants involved in towaway crashes, because "towaway" seems an objective and invariant threshold of crash severity, not dependent on State or local variation of what crashes should be and, in fact, are reported. In the FARS-GES analysis of TTI(d) improvements in Section 2.3, we saw that the towaway criterion was inadvisable because improvements to side structures made cars more damage-resistant and less prone to be towed away. Here, too, we cannot use the towaway criterion, but for the opposite reason: deployment of side air bags may influence a vehicle to be towed when, without the bags, it might have been driven away. Table 3-5 shows the percent of weighted GES crash involvements that were towaways, for the core group of make-models listed in Table 3-1 – without side air bags, with torso bags only, and with torso plus head air bags.

In side impacts, where the deployments take place, the percent of crash-involved cars driven from the scene decreases from 66 percent without side air bags, to 64 percent with torso bags, to 63 percent with both bags, while the percent towed, whether reportedly "due to damage" or "not due to damage," increases from 34 percent without side air bags to 37 percent with both bags. By contrast, in the frontal impacts, where the bags would generally not deploy, the percent towed stays the same. Apparently, people were reluctant to drive away cars with a deployed side air bag because they were eager to move them to a repair facility "as is," or because the deployment made them feel they had experienced a severe crash, and they did not want to drive the car even if it was still operational.[121] Limiting the analysis to towaway crashes would have biased the results in favor of side air bags, because the cars with the air bags had, on the average, less severe towaway crashes, and consequently lower fatality rates per 1,000 towaway crashes. The bias is avoided by simply including all crash involvements on the GES file, towed or not.

120

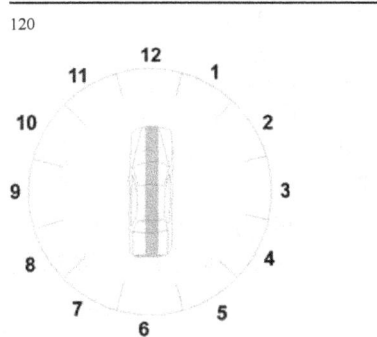

[121] Chidester, A. and Brophy, J. (NHTSA), remarks to the author. Jacobus suggests that some consumers assume their car may be inoperable after a crash with air bag deployment, or they may be cautious and assume their damaged car is not safe to drive. The car is towed away.

TABLE 3-5

PERCENT OF CRASH INVOLVEMENTS THAT ARE TOWAWAYS, WITH AND WITHOUT SIDE AIR BAGS
(1993-2005 GES data)

	Driven Away	Towed Due to Damage	Towed, Not Due to Damage
IN SIDE IMPACTS			
Without side air bags	66	31	3
With torso bags only	64	30	6
With torso bags + head protection	63	33	4
IN FRONTAL IMPACTS			
Without side air bags	55	42	3
With torso bags only	56	38	6
With torso bags + head protection	55	41	4

Basic FARS-GES results Fatality rates – with and without side air bags – are computed in the aggregate for the full "core" group of make-models in Table 3-1, including all the model years specified in that table.[122] They are the make-models that:

- Received some type of side air bags as standard equipment in at least one model year.

- Met FMVSS 214 and were equipped with dual frontal air bags and manual 3-point belts in each of the model years listed for that make-model in Table 3-1, a range of time extending, at most, from model year 1994 to 2003.

- Did not have the same type of side impact protection (namely, no side air bags, torso bags only, or torso bags plus some type of head protection) in every year.

- In any given model year (at least for the model years listed in Table 3-1), all cars of that make-model had the same type of side impact protection, as standard equipment.

Safety belt use has gradually increased from year to year. More recent cars – i.e., the ones with side air bags – have slightly higher belt use. To the extent that belts are effective in nearside impacts (primarily nearside impacts with fixed objects), fatality rates shrink in the more recent cars. A remedy for this possible bias (as in the analyses of Sections 2.3 that included frontal crashes) is to inflate the observed number of FARS fatality cases of occupants protected by safety belts by the inverse of the effectiveness, using the method of NHTSA's evaluation of lives

---

[122] Except Saturn L, the unique make-model that received standard head curtains without torso bags, because there would not be enough cases for meaningful fatality rates with head curtains only.

saved by the FMVSS.[123] For example, if safety belts reduced fatality risk by 5 percent in a multivehicle nearside impact, each fatality case of a belted occupant in that type of crash would be given a weight of

$$1/ (1 - .05) = 1.053 \text{ fatalities}$$

The sum of these weights is an estimate of the number of fatalities that would have occurred if none of the occupants had been protected by safety belts.

TABLE 3-6[124]

MAKE-MODELS WITH <u>STANDARD</u> SIDE AIR BAGS
FATALITIES PER 1,000 NEARSIDE FRONT-SEAT OCCUPANTS IN SIDE IMPACTS
(1993-2005 FARS and GES data; fatalities adjusted upward for safety belt use;
all cars are 214-certified, equipped with dual frontal air bags, and model year 1994-2003;
make-models with the same type of side impact protection in every year are excluded)

	Nearside Fatalities	Nearside Occupants	Fatality Rate	Fatality Reduction
Without side air bags	1,166	276,170	4.22	
With torso bags only	508	145,699	3.49	17 %
With torso bags + head protection	280	101,588	2.75	35 %

Table 3-6 computes the fatality rates for the core make-models. In the model years when they were not equipped with side air bags, side impacts during 1993-2005 involved 276,170 [weighted GES] nearside occupants and resulted in 1,166 [FARS weighted up to zero belt use] fatalities, a rate of 4.22 fatalities per 1,000 exposed occupants. In the model years when they were equipped with torso bags only, the fatality rate was 3.49 (508 fatalities among 145,699 occupants). That is a 17 percent reduction from the rate without side air bags. With torso bags plus some type of head protection (either a torso/head combination bag or a torso bag plus separate head curtain or inflatable tubular structure), the fatality rate was 2.75. That is a cumulative 35 percent reduction from the 4.22 rate without side air bags. The 2.75 fatality rate

---

[123] Safety belts are estimated to reduce fatality risk of nearside occupants by a statistically significant 21 percent in single-vehicle crashes and by a non-significant 5 percent in multivehicle crashes. Kahane, C.J., *Fatality Reduction by Safety Belts for Front-Seat Occupants of Cars and Light Trucks*, NHTSA Technical Report No. DOT HS 809 199, Washington, 2000, p. 30; Kahane, C.J., *Lives Saved by the Federal Motor Vehicle Safety Standards and Other Vehicle Safety Technologies, 1960-2002*, NHTSA Technical Report No. DOT HS 809 833, Washington, 2004, pp. 173-182 and 316-317. Unlike Section 2.3, no adjustment is needed for frontal air bags because every car in this analysis is equipped with frontal air bags. Whereas in Section 2.3, the analyses limited to **multivehicle** nearside impacts were not adjusted for belt effectiveness (because the effect of belts is small and not statistically significant), the analyses of this chapter include single-vehicle as well as multivehicle nearside impacts. Because we must adjust for the 21 percent effect of belts in the single-vehicle crashes, to be consistent, we also adjust for the observed 5 percent effect of belts in the multivehicle crashes.

[124] Tables 3-6 through 3-28 are based on weighted FARS and/or GES data. Those weighted counts are rounded to the nearest integer in the tables, for ease of presentation. However, fatality rates, risk ratios and fatality reductions are based on the original, more accurate, un-rounded counts.

with torso plus head air bags is also 21 percent lower than the 3.49 rate with torso bags only: this is the incremental effectiveness of torso plus head air bags relative to torso bags only.

Statistical significance of these fatality reductions can be tested and confidence bounds computed exactly as in Section 2.3 by treating the effectiveness as a ratio of ratios of FARS and GES statistics. The FARS data were split up into 10 systematic random subsamples, numbered 0 to 9, based on the last digit of the case identification number ST_CASE. GES is a cluster sample of primary sampling units (PSU) that are groups of counties. The GES **PSUs** were randomly split into 10 subgroups of approximately equal size, as follows:

- The PSUs were randomly ordered by issuing each PSU a new PSU number with a SAS random-number generator, and listed in this new order.

- The number of weighted vehicle cases (all vehicle types, all crash types) during 1993-2005 was ascertained for each PSU and cumulated down the list.

- The list was parsed into consecutive (by the new PSU numbers) groups of PSUs, each containing about 10 percent of the cases, and these subgroups were numbered 0 to 9.

The effectiveness of torso bags plus head protection relative to no air bags, more specifically

$$\log r = \log \left[(\text{fatals}_{t+h} / \text{occs}_{t+h}) / (\text{fatals}_{none} / \text{occs}_{none}) \right]$$

is calculated using only the data from FARS subgroup 1 and GES subgroup 1, then recalculated using subgroups 2, and so on, to obtain ten estimates, each based on about 1/10 of the data in Table 3-6. The standard deviation of these ten estimates is .3769. The standard deviation of the corresponding estimate in the full dataset (10 times as many data) is .3769 / $\sqrt{10}$ = .1192. Based on Table 3-6, the estimate for the full dataset is

$$\log \left[(280 / 101,588) / (1,166 / 276,170) \right] = -.4264$$

Because t = −.4264/.1192 = −3.58 is more negative than −1.833, the 5th percentile of a t distribution with 9 degrees of freedom, the fatality reduction for torso plus head air bags is statistically significant at the one-sided .05 level. (In fact, because -3.58 is more negative than -3.25, the 0.5th percentile of a t distribution with 9 degrees of freedom, the reduction is significant at the two-sided .01 level.) The 90 percent confidence bounds[125] for log r are −.4261 $\pm$ 1.833 x .1192 = (−.6449, −.2079). The 90 percent confidence bounds for the fatality-reducing effectiveness (1 − exp[log r]) of torso plus head air bags, relative to no side air bags range from **19 to 48 percent**.

The 17 percent fatality reduction for torso air bags relative to no air bags is also statistically significant at the one-sided .05 level.[126] Confidence bounds extend from 7 to 26 percent. The 21 percent incremental reduction for torso plus head air bags relative torso bags is significant, too.[127] Confidence bounds extend from 3 to 36 percent.

---

[125] -1.833 and +1.833 are the 5th and 95th percentiles of a t-distribution with 9 degrees of freedom.
[126] log r = -.1914, standard error = .0611, t = -3.13.
[127] log r = -.2351, standard error = .1122, t = -2.10.

Effectiveness for drivers vs. passengers Table 3-6a subdivides the data in Table 3-6 by seat position: drivers vs. right-front passengers. Torso plus head air bags are quite effective for both. The fatality reduction is 37 percent for drivers and 31 percent for the passengers. Torso bags alone show 22 percent effectiveness for drivers but just 5 percent for right-front passengers. In Table 3-6a, the reported fatality rates for right-front passengers are nearly twice as high as for drivers. One reason is that they are right-side impacts, which are often severe, such as collisions with fixed objects or with oncoming traffic during a left turn. Another factor could be that uninjured right-front passengers are not recorded in some GES jurisdictions.[128]

TABLE 3-6a

DRIVERS VS. PASSENGERS: MAKE-MODELS WITH <u>STANDARD</u> SIDE AIR BAGS
FATALITIES PER 1,000 NEARSIDE FRONT-SEAT OCCUPANTS IN SIDE IMPACTS
(1993-2005 FARS and GES data; fatalities adjusted upward for safety belt use;
all cars are 214-certified, equipped with dual frontal air bags, and model year 1994-2003;
make-models with the same type of side impact protection in every year are excluded)

	Nearside Fatalities	Nearside Occupants	Fatality Rate	Fatality Reduction
DRIVERS				
Without side air bags	823	225,348	3.65	
With torso bags only	344	120,190	2.86	22 %
With torso bags + head protection	188	81,838	2.29	37 %
RIGHT-FRONT PASSENGERS				
Without side air bags	344	50,822	6.76	
With torso bags only	164	25,509	6.43	5 %
With torso bags + head protection	92	19,749	4.66	31 %

The 37 percent fatality reduction by torso plus head air bags for drivers is exactly the same as McCartt and Kyrychenko's principal estimate (see Section 1.7). Their estimate was also for car drivers, also based on FARS and GES data, but for a somewhat different range of model years and calendar years. The 22 percent reduction for drivers by torso bags is likewise close to their 26 percent estimate.[129]

Although Tables 3-6 and 3-6a are our most comprehensive FARS-GES analyses (highest N of cases), they do not necessarily produce the most accurate effectiveness estimates. Whereas they

---

[128] Because the case vehicles are "in transport" they are all assumed to have drivers, even if no information on the driver is explicitly recorded.
[129] McCartt, A.T. and Kyrychenko, S.Y., *Efficacy of Side Airbags in Reducing Driver Deaths in Driver-Side Car and SUV Collisions*, Insurance Institute for Highway Safety, Arlington, VA, 2006.

include all the core make-models, they do not have matching make-models for different types of side impact protection. For example, because Lincoln Town Car shifted directly from no side air bags to torso/head combination bags, it is included in the rate "without side air bags" but excluded from the rate "with torso bags only." Similarly, because Buick LeSabre did not have head air bags up to 2003, it is included only in the first two rates. In the Sections 3.3 and 3.4 we will develop more focused rates, based on subsets of the core group, that have matching make-models with and without side air bags. Also, analyses based on FARS data alone suggest that effectiveness might be in a somewhat lower range.

FARS analyses of standard side air bags

Purely longitudinal impacts (12:00 or 6:00), where side air bags are unlikely to deploy, can be considered a control group. The effectiveness of side air bags can be estimated from FARS data alone by calculating the change in nearside fatalities relative to these frontal- and rear-impact fatalities. The advantage of working entirely with FARS data is, however, offset by the complexity of adding a control group.

The control group in the analysis of TTI(d) improvements in Section 2.4 was non-occupant fatality cases rather than occupant fatalities in frontals and rear impacts. Frontals are a better control group here because the introduction of side air bags came many years after frontal air bags, whereas the earlier TTI(d) improvements associated with FMVSS 214 coincided or soon followed the introduction of frontal air bags. However, Table 3-8a will present one analysis with a non-occupant control group.[130]

The analysis is based on 1993-2005 FARS data on the core make-models with standard side air bags, listed in Table 3-1.[131] We can use a somewhat more complex definition of impact location based on the variables available in FARS (but not in GES):

- A crash involvement is "nearside" if either the initial impact, or the principal impact, or both are on the near side: 8, 9 or 10:00 for drivers and for 2, 3 or 4:00 right-front passengers. (In the majority of FARS cases, the initial and principal impact locations are the same, most often because there is only a single impact.)

- It is "longitudinal" if the initial and/or the principal impact are at 12:00 or 6:00 while neither impact is a side impact or adjacent to a side impact.[132]

Thus, "nearside" is more inclusive than in the FARS-GES analysis, where the principal impact had to be on the near side. A car that is hit first in the side and then in the front, or vice-versa, would be a nearside impact for the occupant on that side because, either way, the bags could

---

[130] Dainius Dalmotas, in his peer review of this report, recommended the non-occupant control group for the FARS analyses of Chapter 2, but did not make a similar recommendation for this chapter.
[131] In addition to the model years in the text of Table 3-1 (the years used in the GES-FARS analysis), we have added a few years, such as 1999 Volkswagen Jetta, where the type of side impact protection changed in mid-year, but is clearly decodable from the VIN, as described in the footnotes to Table 3-1.
[132] For both of these definitions, if one of the impact areas is unknown or not reported, rely on the area that is reported. "A side impact" includes 2-4:00 and 8-10:00 (regardless of where the occupant is sitting) and "adjacent to a side impact" includes 1:00, 5:00, 7:00 and 11:00. All these are excluded, regardless of the occupant's seat position, because they could deploy side air bags.

deploy. "Longitudinal" is limited to cars without any lateral damage: a real control group where the bags are unlikely to deploy. Fatality cases of belted occupants are weighted higher than 1 to estimate the number of fatalities that would have occurred if none of the occupants had been protected by safety belts.[133]

Contingency table analysis Table 3-7 computes the ratio of nearside to purely longitudinal fatalities for the core make-models. The cars that were not equipped with side air bags experienced 1,304 nearside[134] and 2,365 longitudinal fatalities during 1993-2005, a risk ratio of .551. With torso bags only, the risk ratio was .474 (580 nearside vs. 1,224 longitudinal fatalities). That is a 14 percent reduction from the risk ratio without side air bags. With torso bags plus some type of head air bag, the risk ratio was .409: a cumulative 26 percent reduction from the rate without side air bags. It is also an incremental 14 percent reduction relative to torso bags only.

TABLE 3-7

MAKE-MODELS WITH <u>STANDARD</u> SIDE AIR BAGS
NEARSIDE VS. LONGITUDINAL FATALITIES
(1993-2005 FARS; fatalities adjusted upward for safety belt use;
all cars are 214-certified, equipped with dual frontal air bags, and model year 1994-2003;
make-models with the same type of side impact protection in every year are excluded)

	Nearside Fatalities	12:00 or 6:00 Fatalities	Risk Ratio	Nearside Reduction
Without side air bags	1,304	2,365	.551	
With torso bags only	580	1,224	.474	14 %
With torso bags + head protection	324	792	.409	26 %

Statistical significance of the fatality reductions in Table 3-7 cannot be tested with simple 2x2 chi-squares of the fatality counts in any two rows (e.g., the first and third rows for the effect of torso bags plus head protection relative to no air bags), because they are weighted counts, not actual numbers of cases. Belted FARS cases have been weighted more than 1 to adjust for the effect of belt use. Instead, the FARS data were split up into 10 systematic random subsamples based on the last digit of the case identification number ST_CASE, and significance was tested by observing the variation of the estimate across subsamples. The effectiveness of torso plus head air bags relative to no air bags, more specifically

$$\log r = \log \left[(\text{nearside}_{t+h} / \text{frontal}_{t+h}) / (\text{nearside}_{none} / \text{frontal}_{none}) \right]$$

---

[133] Kahane (2004), pp. 173-182 and 316-317. Pp. 316-317 estimate belt effectiveness for various types of crashes, based on Kahane (2000), p. 30.

[134] The counts of nearside fatalities in Table 3-7 are slightly larger than in Table 3-6, primarily because of the more inclusive definition of "nearside," and to a lesser extent because of the additional vehicles such as 1999 Volkswagen Jetta.

is calculated using only the data from FARS subgroup 1, then recalculated using subgroup 2, and so on, to obtain ten estimates, each based on about 1/10 of the data in Table 3-7. The standard deviation of these ten estimates is .3531. The standard deviation of the corresponding estimate in the full dataset (10 times as many data) is .3531 / √10 = .1116. Based on Table 3-7, the estimate for the full dataset is

$$\log [(324 / 792) / (1,304 / 2,365)] = -.2973$$

Because t = –.2973 / .1116 = –2.66 is more negative than –1.833 (the 5th percentile of a t-distribution with 9 degrees of freedom), the 26 percent fatality reduction for torso plus head air bags is statistically significant. The 90 percent confidence bounds for fatality reduction are 1 - exp (–.2973 ± 1.833 x .1116), or 9 to 39 percent. However, the 14 percent fatality reduction for torso bags is not statistically significant.[135] Neither is the 14 percent incremental fatality reduction for torso plus head air bags relative to torso bags.[136]

In other words, the basic FARS-GES analysis of Table 3-6 generated a somewhat higher estimate of fatality reduction for torso bags plus head protection (35 vs. 26 percent) and a slightly higher estimate for torso air bags as well (17 vs. 14 percent).

Logistic regression allows an adjustment for some imbalances and trends in the basic data. Some make-models are underrepresented, or not represented at all with certain types of side impact protection (e.g., Lincoln Town Car was never offered with torso bags only). The tendency to get into side impacts as opposed to longitudinal impacts could vary with the age of the car, the age and gender of the driver, or the geographic location – parameters that are known from the FARS data and can be controlled by regression.

The setup for the logistic regression resembles the one in Section 2.4. In other words, the data points in the regression are the FARS occupant fatality cases (possibly given a weight factor greater than 1 if belted). The dependent variable, NEARSIDE equals 1 for nearside fatalities, 2 for purely longitudinal impacts. There are two key independent variables:

- BOTHBAGS = 1 for cars with torso bags plus head air bags (including combination bags), = 0 for all other cars.

- TORSO_ONLY = 1 for cars with torso bags only, = 0 for all other cars.

The other independent variables include:

- A set of dichotomous variables representing the make-models in Table 3-1 (sometimes aggregated into groups when N is relatively small – e.g., all Audis, all Volvos).

- A set of dichotomous variables representing the vehicle or object struck (a car, an LTV, a heavy truck, a fixed object, or other/unknown/3+ vehicle crash).

- The occupant's age, expressed as AGEGE18, the age minus 18 (but set to zero if the occupant was 12-18 years old). This recognizes that fatality risk given a similar physical

---

[135] log r = -.1515, standard error = .0896, t = -1.69.
[136] log r = -.1458, standard error = .1142, t = -1.28.

insult rises steadily from age approximately 18 onward, but not from 12 to 18.[137] Cases with unreported age are excluded; so are right-front passengers younger than 12 years.

- Gender, expressed as FEMALE (= 1 for females, 0 for males).

- Seat position, expressed as RF (= 1 for right-front passengers, 0 for drivers)

- Vehicle age (VEHAGE)

- Speed limit, expressed as SPDLIM55 (= 1 if it is 55+, 0 if it is 50 or less; when the various vehicles are on different roads, this is the highest speed limit for any of the roads)

- Time of day, expressed as NITE (= 1 if 7:00 p.m. – 5:59 a.m., 0 if 6:00 a.m. – 6:59 p.m.)

- RURAL (= 1 if rural, 0 if urban)

- FREEWAY (= 1 if limited access divided highway, 0 otherwise)

- WETROAD (= 1 if surface condition was wet, 0 otherwise)

- HIFAT_ST (= 1 for the 26 States that have had the highest fatality rates per registered vehicle; 0 for the others)[138]

The coefficient for BOTHBAGS is -.408. In other words, the combination of torso and head air bags is associated with a $1 - \exp(-.408) = 34$ percent reduction in nearside relative to longitudinal-impact fatality risk. The coefficient for TORSO_ONLY is -.306, corresponding to a $1 - \exp(-.306) = 26$ percent fatality reduction for the torso bags alone. These estimates are not inconsistent with the contingency table analysis, but the estimate for torso plus head air bags is especially close to the basic FARS-GES result (35 percent in Table 3-5).

FARS analyses of standard plus optional side air bags

With FARS data, information on a car's make-model and its side impact protection can nearly always be derived from the VIN. The core make-models in Table 3-1 may be supplemented with the models in Table 3-2 that had optional, but VIN-decodable side air bags in certain years, expanding the N of cases available. In these models there is, furthermore, some balance between the cars with the various alternative types of side impact protection: side air bags are present in at least 1/6 and at most 2/3 of the cars. The side air bags may be entirely the buyer's option, unrelated to other equipment on the car, or they may be standard equipment on certain sub-series and unavailable on others.

Contingency table analysis Table 3-8 computes the ratio of nearside to purely longitudinal fatalities. The inclusion of the make-models with optional air bags increases the N by about a third over Table 3-7. Cars without side air bags experienced a risk ratio of .493. With torso bags only, the risk ratio was .484. That is little change from the risk ratio without side air bags. With torso bags plus some type of head air bag, the risk ratio was .390: a cumulative 21 percent reduction from the rate without side air bags. It is also an incremental 20 percent reduction relative to torso bags only. The cumulative 21 percent fatality reduction for torso plus head bags

---

[137] Evans, L., *Traffic Safety and the Driver,* Van Nostrand Reinhold, New York, 1991, pp. 25-28.
[138] Kahane, C.J., *Vehicle Weight, Fatality Risk and Crash Compatibility of Model Year 1991-99 Passenger Cars and Light Trucks,* NHTSA Technical Report No. DOT HS 809 662, Washington, 2003, pp. 24-26.

relative to no side air bags is statistically significant. The 90 percent confidence bounds for fatality-reducing effectiveness range from 12 to 29 percent.[139]

TABLE 3-8

MAKE-MODELS WITH <u>STANDARD OR OPTIONAL</u> SIDE AIR BAGS
NEARSIDE VS. LONGITUDINAL FATALITIES
(1993-2005 FARS; fatalities adjusted upward for safety belt use;
all cars are 214-certified, equipped with dual frontal air bags, and model year 1994-2003;
make-models with the same type of side impact protection in every year are excluded)

	Nearside Fatalities	12:00 or 6:00 Fatalities	Risk Ratio	Nearside Reduction
Without side air bags	1,991	4,039	.493	
With torso bags only	727	1,501	.484	2 %
With torso bags + head protection	448	1,150	.390	21 %

The combined effect of torso plus head air bags, 21 percent, is slightly lower than the results in the preceding analyses, but here, for the first time, head protection accounts for all of the effect. However, a comparison of Tables 3-7 and 3-8 shows the risk ratio with torso bags barely increased from .474 to .484. Rather, the risk ratio without side air bags decreased from .551 to .493 and the ratio with torso plus head air bags decreased from .409 to .390; thus, torso bags became less effective relative to the other two groups. These phenomena could be due to a different make-model mix with and without air bags (something we can control for with logistic regression), or a tendency of drivers especially prone to side impacts (e.g., female drivers) to self-select torso air bags when they are an option.

Logistic regression Except for the addition of some dichotomous variables to represent the make-models in Table 3-2 with optional air bags, the logistic regression is identical to the preceding one. Here, the coefficient for BOTHBAGS is -.213. In other words, the combination of torso and head air bags is associated with a $1 - \exp(-.213) = 19$ percent fatality reduction. The coefficient for TORSO_ONLY is -.136, corresponding to a $1 - \exp(-.136) = 13$ percent fatality reduction for the torso bags alone. The combined effect is slightly lower than in all the previous analyses. The adjustments made possible by logistic regression have brought the effect for torso bags only into line with the preceding results.

Contingency table analysis with a non-occupant control group Table 3-8a computes the ratio of nearside to non-occupant fatalities. As in Section 2.4, these "non-occupant fatalities" are the pedestrians, bicyclists and other non-motorists struck and fatally injured by cars, in this case, by the cars of the make-models and model-year ranges defined in Tables 3-1 and 3-2. The analysis considers the ratio of fatalities who were nearside occupants **in** the vehicles to the fatalities among pedestrians, bicyclists and other non-motorists who were struck **by** the vehicles [in other

---

[139] Same method as with Table 3-7; log r = -.2356, standard error = .0599, t = -3.93, df = 9.

crashes]. Cars without side air bags experienced a risk ratio of 1.671.[140] With torso bags only, the risk ratio was 1.699. As in Table 3-8, that is little change from the risk ratio without side air bags. With torso plus head air bags, the risk ratio was 1.317: a 21 percent reduction from the rate without side air bags. That estimate is identical to the 21 percent reduction, in Table 3-8, with the control group of occupant fatalities in purely longitudinal crashes.[141] We conclude that, for analyses of side air bags, the results are about the same with the longitudinal and the non-occupant control groups. Because occupant fatalities in longitudinal impacts are 3-4 times as numerous as non-occupant fatalities, they are the preferable control group, because they contribute more N to the analyses.

TABLE 3-8a

MAKE-MODELS WITH <u>STANDARD OR OPTIONAL</u> SIDE AIR BAGS
NEARSIDE VS. NON-OCCUPANT FATALITIES
(1993-2005 FARS; fatalities adjusted upward for safety belt use;
all cars are 214-certified, equipped with dual frontal air bags, and model year 1994-2003;
make-models with the same type of side impact protection in every year are excluded)

	Nearside Fatalities	Non-Occupant Fatalities	Risk Ratio	Nearside Reduction
Without side air bags	1,991	1,192	1.671	
With torso bags only	727	428	1.699	− 2 %
With torso bags + head protection	448	340	1.317	21 %

Analyses for some groups of crashes that should not be affected by side air bags

Side air bags are not necessarily the only factor that reduced fatality risk in nearside impacts in recent years. If fatality rates have generally declined over time or are lower in newer/more recent cars, such trends might be contributing some of the fatality reduction seen in the preceding analyses. Therefore, it is useful to look at the trend in fatality rates over time in groups of cars or crashes where side air bags are not a factor.

FARS-GES nearside fatality rates in cars that never had side air bags Table 3-3 lists 19 make-models that did not have side air bags until at least 2003. If nearside fatality rates decreased in

---

[140] Counts of nearside fatalities are identical in Tables 3-8 and 3-8a. Unlike Table 3-8, where FARS counts of occupant fatalities in longitudinal impacts are inflated to account for belt use, the non-occupant numbers in Table 3-8a are the actual FARS counts. A special technique is needed for the make-models that had different air bags in the driver and right-front seat positions (Buick Century and Regal, Cadillac Seville, Chevrolet Caprice and Monte Carlo – see Tables 3-1 and 3-2). Given a customary 3:1 ratio of drivers to right-front passengers in crash-involved cars, 75 percent of the non-occupant fatalities in these make-models are allocated to analyses of drivers' fatality risk and 25 percent to right-front passengers'.

[141] Similarly, when the analysis is limited to make-models with standard side air bags, Table 3-7 estimated a 14 percent reduction for torso bags only and 26 percent with torso plus head air bags, with occupant fatalities in longitudinal impacts as the control group. With non-occupant fatalities as the control group, these estimates are 9 and 29 percent, respectively.

these cars, over time, before 2003, it obviously wouldn't have been due to side air bags. The span of model years for each make-model in Table 3-3 is split into "early," "mid" and "late" years – e.g., a span from 1994 to 2003 is split into 1994-1996, 1997-1999 and 2000-2003. Table 3-9 shows the nearside fatality rates were quite similar in the early, middle and late model years for the cars that did not receive side air bags, and it shows no evidence of a trend to lower fatality rates over time.

TABLE 3-9

MAKE-MODELS THAT NEVER RECEIVED SIDE AIR BAGS
FATALITIES PER 1,000 NEARSIDE FRONT-SEAT OCCUPANTS IN SIDE IMPACTS
(1993-2005 FARS and GES data; fatalities adjusted upward for safety belt use;
all cars are 214-certified, equipped with dual frontal air bags, and model year 1994-2003)

	Nearside Fatalities	Nearside Occupants	Fatality Rate	Fatality Reduction
Early model years	1,720	395,465	4.35	
Mid model years	1,181	282,740	4.18	4 %
Late model years	769	180,899	4.25	2 %

FARS-GES frontal fatality rates in models that received side air bags as standard equipment
Side air bags ought to have little effect in frontal crashes, except to the extent that some of the crashes might be oblique enough for bags to deploy and protect occupants. Nevertheless, Table 3-10 shows that frontal fatality rates were 5 percent lower in the cars with torso air bags and 17 percent lower with torso bags plus head protection than in the cars with no side air bags. Although these "effects" are not as large as the fatality reductions for nearside occupants (Table 3-6), the 17 percent reduction is statistically significant.[142] Unlike Table 3-9, the findings in Table 3-10 cannot just be ignored. Nevertheless, it is important to note that any bias suggested by the results in Table 3-10 would apply only to FARS-GES analyses, because in the FARS-only analyses, fatality risk in nearside impacts is computed relative to risk in longitudinal (primarily frontal) impacts. The FARS analyses automatically control for any long-term absolute change in frontal fatality risk, because they only measure how much the reduction in nearside fatalities exceeds the reduction in frontals.

---

[142] By the method discussed with Table 3-6, log r = -.1882, standard error = .0705, t = -2.67, df = 9.

TABLE 3-10

MAKE-MODELS WITH <u>STANDARD</u> SIDE AIR BAGS
FATALITIES PER 1,000 FRONT-SEAT OCCUPANTS IN <u>FRONTAL</u> IMPACTS
(1993-2005 FARS and GES data; fatalities adjusted upward for safety belt use;
all cars are 214-certified, equipped with dual frontal air bags, and model year 1994-2003;
make-models with the same type of side impact protection in every year are excluded)

	Frontal Fatalities	N of Occupants	Fatality Rate	Fatality Reduction
Without side air bags	2,786	663,030	4.20	
With torso bags only	1,335	333,785	4.00	5 %
With torso bags + head protection	895	257,195	3.48	17 %

FARS analysis of cars that never had side air bags Table 3-11 shows the risk ratio of nearside to longitudinal fatalities decreased by a cumulative 12 percent from the early to the late model years for the cars that did not receive side air bags.

TABLE 3-11

MAKE-MODELS THAT <u>NEVER RECEIVED</u> SIDE AIR BAGS
NEARSIDE VS. LONGITUDINAL FATALITIES
(1993-2005 FARS; fatalities adjusted upward for safety belt use;
all cars are 214-certified, equipped with dual frontal air bags, and model year 1994-2003)

	Nearside Fatalities	12:00 or 6:00 Fatalities	Risk Ratio	Nearside Reduction
Early model years	1,874	3,137	.597	
Mid model years	1,291	2,257	.572	4 %
Late model years	832	1,584	.525	12 %

The cumulative 12 percent reduction is statistically significant.[143] But it is not even half as large as the 26 percent fatality reduction for torso plus head air bags in the cars that received them as standard equipment (Table 3-7). Furthermore, the three risk ratios in Table 3-11 (i.e., without side air bags), .597, .572 and .525 straddle the .551 risk ratio without side air bags in Table 3-7 and are far above the .409 risk ratio in Table 3-7 with torso plus head air bags.

The three control-group analyses did not show a consistent trend. Table 3-9 showed close to zero cumulative effect; Table 3-10, a fairly sizable effect (relevant only to FARS-GES analyses); and Table 3-11, something in between. The results are not sufficiently clear-cut to establish specific adjustment factors that should be deducted from our previous effectiveness estimates,

---

[143] By the method discussed with Table 3-7, log r = -.1282, standard error = .0612, t = -2.09, df = 9.

but they do suggest that estimates should be viewed with some caution. For example, where different analyses have generated a range of effectiveness, the best answer is probably somewhat lower than the median of that range.

3.3 Fatality reduction by torso air bags in nearside impacts

More accurate estimates of the effectiveness of torso air bags may be obtained by limiting the data to the make-models that shifted from no side air bags to standard torso air bags, or that offered a choice of no air bags or torso air bags. For example, in Table 3-1, we would:

- Include Buick LeSabre for the entire time span (1997-2002), because it had no side air bags in 1997-1999 and torso bags in 2000-2002.

- Exclude Lincoln Town Car because it was never equipped with torso air bags only.

- Include Volkswagen Passat, but only for model years 1995-2000, when it was equipped with no side air bags or torso air bags only.

- Exclude BMW 500 because it was never FMVSS 214-certified without side air bags.

In Table 3-2, we would

- Include Cadillac Catera for the entire time span, because it offered a choice of no side air bags or torso air bags.

- Exclude Chrysler PT Cruiser, because it never offered the torso air bag alone.

- Include Nissan Maxima, but only for model years 1998-1999, when it offered a choice of no side air bags or torso air bags.

These exclusions provide datasets that comprise exactly the same list of make-models with no side air bags and with torso air bags (although not necessarily in the same proportions), allowing a more direct comparison of fatality risk before and after the introduction of torso air bags.

FARS-GES analysis Table 3-12 computes fatality rates for drivers and right-front passengers who are nearside occupants in side impact crashes. As in Section 3.2, the numerator of the fatality rates is FARS cases, with belted cases given a weight greater than 1 to adjust for the effect of belt use; the denominator is weighted GES cases. For the make-models that shifted from no side air bags to standard torso bags, 207,218 nearside occupants of cars without the bags experienced 853 fatalities, a rate of 4.12 fatalities per 1,000 exposed occupants. With torso bags, the fatality rate was 3.50. That is a statistically significant 15 percent reduction from the rate without side air bags. The 90 percent confidence bounds for fatality reduction are 8 to 22 percent.[144]

---

[144] By the method discussed with Table 3-6, log r = -.1632, standard error = .0439, t = -3.72, df = 9.

TABLE 3-12

MAKE-MODELS SHIFTING FROM <u>NO</u> SIDE AIR BAGS TO <u>STANDARD</u> TORSO BAGS
FATALITIES PER 1,000 NEARSIDE FRONT-SEAT OCCUPANTS IN SIDE IMPACTS
(1993-2005 FARS and GES data; fatalities adjusted upward for safety belt use;
all cars are 214-certified, equipped with dual frontal air bags, and model year 1994-2003)

	Nearside Fatalities	Nearside Occupants	Fatality Rate	Fatality Reduction
Without side air bags	853	207,218	4.12	
With torso bags only	497	142,111	3.50	15 %

Table 3-12 can be compared to Table 3-6 to see how much was included or excluded. Almost all the cases with torso bags in the basic analysis were retained (497 of 508 fatalities and 142,111 of 145,699 weighted occupants). But many cases without side air bags are gone (down from 1,166 to 853 fatalities and from 276,170 to 207,218 weighted occupants), because the many models that shifted directly from no air bags to torso bags plus head protection were deleted. Still, the N of cases without side air bags remains substantially larger than the N with torso bags.

As discussed in Section 1.4, the introduction of torso air bags often came within a year or two, or even coincided with certification to the head-impact upgrade of FMVSS 201 (without head air bags). Specifically, in Table 3-12, only 1 percent of the cars without side air bags are FMVSS 201-certified, but 37 percent of the cars with torso bags only are 201-certified. With more detailed crash data that would not be an issue, because torso bags are primarily designed to mitigate torso injury, whereas the energy-absorbing materials used in certifying to FMVSS 201 would mitigate head injury. Because the basic FARS data do not identify the body region injured, such distinctions are impossible and some of the overall fatality reduction in Table 3-12 could be due to FMVSS 201, not the torso bags. However, the vehicle cases in Table 3-12 can be subdivided according to whether

- The car was 201-certified.

- The make-model was 201-certified before, simultaneously, or after torso bags became standard equipment.

Two subgroups comprise much of the data:

1. Make-models that certified to FMVSS 201 in the same year they received torso bags.

2. Pre-FMVSS 201 cars in make-models that did not certify to FMVSS 201 until at least two years after they received torso bags.

In the first subgroup, the analysis of Table 3-12 really measures the combined effect of torso bags and FMVSS 201 compliance (by energy-absorbing materials). In the second subgroup, it measures purely the effect of torso bags. The observed fatality reduction in subgroup 1 (for torso bags plus FMVSS 201) is −5 percent, while the reduction in subgroup 2 (for torso bags only) is

12 percent. The direction of these results hardly suggests that the fatality reduction claimed for torso bags in Table 3-12 is instead primarily due to FMVSS 201. However, they are based on limited data. We cannot yet exclude the possibility that some portion, likely not a large one, of the fatality reduction claimed for torso bags in Table 3-12 is actually due to FMVSS 201.

FARS analysis of standard torso bags Table 3-13 computes the ratio of nearside to purely longitudinal (12:00 or 6:00) fatalities for the make-models that shifted from no side air bags to standard torso bags. Cars without side air bags experienced 946 nearside and 1,668 longitudinal fatalities during 1993-2005, a risk ratio of .567. With torso bags, the risk ratio for these same make-models was .472. That is a statistically significant 17 percent reduction. Confidence bounds range from 2 to 29 percent.[145] Also, in Table 3-13, the possible confounding effect of FMVSS 201 may be less of an issue because energy-absorbing materials can mitigate head injuries in frontal and rear impacts (the control group) as well as in side impacts; the effect of FMVSS 201 in the side impacts might be offset by a possibly similar effect in the control group.

TABLE 3-13

MAKE-MODELS SHIFTING FROM <u>NO</u> SIDE AIR BAGS TO <u>STANDARD</u> TORSO BAGS
NEARSIDE VS. LONGITUDINAL FATALITIES
(1993-2005 FARS; fatalities adjusted upward for safety belt use;
all cars are 214-certified, equipped with dual frontal air bags, and model year 1994-2003)

	Nearside Fatalities	12:00 or 6:00 Fatalities	Risk Ratio	Nearside Reduction
Without side air bags	946	1,668	.567	
With torso bags only	562	1,191	.472	17 %

FARS analysis of standard plus optional torso air bags Table 3-14 supplements the preceding data with make-models that offered a choice between no side air bags or torso bags only (see Table 3-2). With these additional cases, the fatality reduction for torso bags is a non-significant 5 percent.[146]

---

[145] By the method discussed with Table 3-7, log r = -.1838, standard error = .0875, t = -2.10, df = 9.
[146] By the method discussed with Table 3-7, log r = -.0499, standard error = .0817, t = - .61, df = 9.

TABLE 3-14

MAKE-MODELS SHIFTING FROM <u>NO</u> SIDE AIR BAGS TO <u>STANDARD</u> TORSO BAGS
OR OFFERING A CHOICE BETWEEN NO AIR BAGS AND TORSO BAGS
NEARSIDE VS. LONGITUDINAL FATALITIES
(1993-2005 FARS; fatalities adjusted upward for safety belt use;
all cars are 214-certified, equipped with dual frontal air bags, and model year 1994-2003)

	Nearside Fatalities	12:00 or 6:00 Fatalities	Risk Ratio	Nearside Reduction
Without side air bags	1,335	2,621	.509	
With torso bags only	712	1,471	.485	5 %

3.4 Fatality reduction by torso bags plus head protection in nearside impacts

As in Section 3.3, we hope to obtain more accurate effectiveness estimates by comparing fatality risk for the same list of make-models, with and without head air bags. But the analysis is more complicated for head air bags than torso air bags, for several reasons:

- Whereas all models that shifted from no air bags to standard torso bags did so directly, that is not true for head air bags:
    - Some models, such as Lincoln Town Car, shifted directly from no air bags to torso bags plus head protection.
    - Others, such as Volkswagen Passat, shifted initially from no air bags to torso bags only, and finally to torso bags plus head protection.

- For the models that shifted directly to torso bags plus head protection, we can estimate the combined effect of both bags, but cannot isolate how much of the reduction is due to torso bags, and how much to head air bags.

- There are two principal types of head protection: combination torso/head air bags and separate head curtains (including inflatable tubular structures in BMW). We would like separate results for each type, to the extent possible with the limited data currently available.

Overall effectiveness of torso bags plus head protection relative to no side air bags

Let us first compile a list of make-models from Table 3-1 that initially had no side air bags (but were FMVSS 214-certified) and eventually received torso plus head air bags as standard equipment – either directly (Lincoln Town Car) or with a transitional phase of torso bags only (Volkswagen Passat). However, we will limit the data to the model years with no air bags or with torso bags plus head protection – e.g., the full 1994-2003 time span for Lincoln Town Car, but only 1995-1997 and 2001-2003 for Volkswagen Passat.

FARS-GES analysis Table 3-15 computes fatality rates for drivers and right-front passengers who are nearside occupants in side impact crashes.

TABLE 3-15

MAKE-MODELS THAT ONCE HAD <u>NO</u> SIDE AIR BAGS
AND NOW HAVE <u>STANDARD</u> TORSO BAGS PLUS HEAD PROTECTION
FATALITIES PER 1,000 NEARSIDE FRONT-SEAT OCCUPANTS IN SIDE IMPACTS
(1993-2005 FARS and GES data; fatalities adjusted upward for safety belt use;
all cars are 214-certified, equipped with dual frontal air bags, and model year 1994-2003)

	Nearside Fatalities	Nearside Occupants	Fatality Rate	Fatality Reduction
Without side air bags	660	163,018	4.05	
With torso bags + head protection	265	95,349	2.78	31 %

When these make-models did not have side air bags, the fatality rate was 4.05 per 1,000 nearside occupants. When the same models were equipped with torso plus head air bags, the fatality rate was 2.78. That is a statistically significant 31 percent reduction from the rate without side air bags. The 90 percent confidence bounds for fatality reduction are 11 to 47 percent.[147]

Section 1.4 listed alternative strategies for certifying to FMVSS 201 in make-models that eventually received head air bags. Manufacturers could:

- Initially certify to FMVSS 201, using energy-absorbing materials where needed, and add the air bags in a later year.

- Certify to 201 at the same time they installed air bags, if necessary complementing the air bags with energy-absorbing materials.

- Certify to 201 a year or more after installing head air bags. The delay in certification could be due to:

  o A technicality: the manufacturer was ahead of the phase-in schedule and had no need to certify that make-model, even though it could have complied, or

  o Energy-absorbing materials subsequently added to fully meet FMVSS 201.

Specifically, in Table 3-15, none of the cars without side air bags are FMVSS 201-certified. Among the cars with torso and head air bags:

- 22 percent are in make-models that had certified to 201 before they received head air bags (namely, during years when they were equipped with torso bags only).

- 46 percent are in 201-certified cars of make-models that certified to 201 in the year they received head air bags or in a subsequent year.

- 32 percent are in cars that were not 201-certified (and perhaps many of these cars could have met FMVSS 201, but simply were not certified).

---

[147] By the method discussed with Table 3-6, log r = -.3753, standard error = .1418, t = -2.65, df = 9.

In other words, most of the cars with head air bags, at least 68 percent, are 201-certified and/or could have met FMVSS 201. In that sense, Table 3-15 measures the effect not only of head air bags but also of any other modifications in response to FMVSS 201. However, only 22 percent of the cars were from make-models that had initially certified to 201 with energy-absorbing materials, without head air bags. By contrast, 78 percent of the sample are from models that certified to 201 simultaneous with receiving head air bags, or later. NHTSA's cost analyses of FMVSS 201, based on a survey of 14 specimen vehicles, found little change in the energy-absorbing materials at the head-impact targets when vehicles initially certified to FMVSS 201 with head air bags.[148] Thus, even though Table 3-15 measures the combined effect of head air bags and any other modifications in response to FMVSS 201, the "any other modifications" may well be negligible on most of these cars.

FARS analysis of standard torso bags plus head protection Table 3-16 computes the ratio of nearside to purely longitudinal (12:00 or 6:00) fatalities. When these make-models did not have side air bags, the fatality risk ratio was .531. With torso bags plus head protection, the fatality risk ratio was .424. That is a 20 percent reduction. It is not statistically significant.[149] Also, in Table 3-16, as in Table 3-13, the possible confounding effect of FMVSS 201 may be less of an issue because energy-absorbing materials can mitigate head injuries in frontal and rear impacts (the control group) as well as in side impacts.

TABLE 3-16

MAKE-MODELS THAT ONCE HAD <u>NO</u> SIDE AIR BAGS
AND NOW HAVE <u>STANDARD</u> TORSO BAGS PLUS HEAD PROTECTION
NEARSIDE VS. LONGITUDINAL FATALITIES
(1993-2005 FARS; fatalities adjusted upward for safety belt use;
all cars are 214-certified, equipped with dual frontal air bags, and model year 1994-2003)

	Nearside Fatalities	12:00 or 6:00 Fatalities	Risk Ratio	Nearside Reduction
Without side air bags	746	1,404	.531	
With torso + head protection	309	728	.424	20 %

FARS analysis of standard + optional torso plus head air bags Let us supplement the data in Table 3-16 with the make-models in Table 3-2 that offered a choice between no side air bags or torso bags plus head protection. For example, we would include:

- Chrysler PT Cruiser for the entire time span in Table 3-2 (2001-2003), when it offered a choice of no side air bags or torso/head combination bags.

---

[148] Ludtke, N.F., Osen, W., Gladstone, R. and Lieberman, W, *Perform Cost and Weight Analysis, Head Protection Air Bag Systems, FMVSS 201*, NHTSA Technical Report No. DOT HS 809 842, Washington, 2004, pp. 3-47 – 3-54; i.e., there were, on the whole, no substantial cost increases (or decreases) in the components that house the energy-absorbing materials.

[149] By the method discussed with Table 3-7, log r = -.2254, standard error = .1510, t = -1.49, df = 9.

- Nissan Maxima, but only for model years 2000-2003, when it offered a choice of no side air bags or combination bags.

Table 3-17 computes the ratio of nearside to purely longitudinal fatalities. Without side air bags, the fatality risk ratio was .491. For the same make-models with torso bags plus head protection, the ratio was .396. That is a statistically significant 19 percent reduction from the rate without side air bags.[150] The 90 percent confidence bounds for fatality-reducing effectiveness range from 4 to 32 percent.

TABLE 3-17

MAKE-MODELS THAT ONCE HAD <u>NO</u> SIDE AIR BAGS
AND NOW HAVE <u>STANDARD</u> TORSO BAGS PLUS HEAD PROTECTION,
OR OFFERING A CHOICE BETWEEN NO AIR BAGS AND TORSO + HEAD AIR BAGS
NEARSIDE VS. LONGITUDINAL FATALITIES
(1993-2005 FARS; fatalities adjusted upward for safety belt use;
all cars are 214-certified, equipped with dual frontal air bags, and model year 1994-2003)

	Nearside Fatalities	12:00 or 6:00 Fatalities	Risk Ratio	Nearside Reduction
Without side air bags	1,051	2,140	.491	
With torso + head protection	433	1,092	.396	19 %

Combination torso/head air bags vs. separate head curtains (with torso bags)

The data in Tables 3-15, 3-16 and 3-17 may be subdivided into the models that received combination torso/head air bags and the models that received separate head curtains (including inflatable tubular structures) with torso bags. The currently available data do not support a conclusion that one type of head protection is more effective than the other. Table 3-18, based on the FARS-GES analysis, shows a 28 percent fatality reduction for combination bags (relative to no side air bags) and a 29 percent reduction for head curtains plus torso bags. But the two analyses based on FARS data alone yield higher point estimates of effectiveness for the combination bags: Table 3-19 (standard air bags only) shows 28 percent fatality reduction for combination bags and 14 percent for head curtains; Table 3-20 (standard plus optional bags) finds 26 and 9 percent, respectively.[151] We may assume, until more data become available, that both types of head protection are about equally effective in nearside impacts.

---

[150] By the method discussed with Table 3-7, log r = -.2153, standard error = .0926, t = -2.32, df = 9.

[151] A few models shifted from no bags to combination bags to, eventually, head curtains plus torso bags. For these make-models, the vehicles without side air bags are a predecessor to both the vehicles with combination bags and the vehicles with side curtains. Thus, the data in Tables 3-18 – 3-21 for combination bags and side curtains add up exactly (within rounding) to the data for torso bags plus head protection in corresponding Tables 3-15 – 3-17, but the data in Tables 3-18 – 3-21 without side air bags add up to slightly more than the data without side air bags in corresponding Tables 3-15 – 3-17.

TABLE 3-18

MAKE-MODELS THAT ONCE HAD <u>NO</u> SIDE AIR BAGS
AND NOW HAVE <u>STANDARD</u> TORSO BAGS PLUS HEAD PROTECTION
FATALITIES PER 1,000 NEARSIDE FRONT-SEAT OCCUPANTS IN SIDE IMPACTS
(1993-2005 FARS and GES data; fatalities adjusted upward for safety belt use;
all cars are 214-certified, equipped with dual frontal air bags, and model year 1994-2003)

	Nearside Fatalities	Nearside Occupants	Fatality Rate	Fatality Reduction
Make/models that shifted to torso/head combination bags				
Without side air bags	418	101,088	4.13	
With torso/head combination bags	132	44,564	2.96	28 %
Make/models that shifted to head curtains (plus torso bags)				
Without side air bags	264	70,887	3.73	
With head curtains (+ torso bags)	134	50,785	2.63	29 %

TABLE 3-19

MAKE-MODELS THAT ONCE HAD <u>NO</u> SIDE AIR BAGS
AND NOW HAVE <u>STANDARD</u> TORSO BAGS PLUS HEAD PROTECTION
NEARSIDE VS. LONGITUDINAL FATALITIES
(1993-2005 FARS; fatalities adjusted upward for safety belt use;
all cars are 214-certified, equipped with dual frontal air bags, and model year 1994-2003)

	Nearside Fatalities	12:00 or 6:00 Fatalities	Risk Ratio	Nearside Reduction
Make/models that shifted to torso/head combination bags				
Without side air bags	479	957	.500	
With torso/head combination bags	144	398	.362	28 %
Make/models that shifted to head curtains (plus torso bags)				
Without side air bags	292	503	.580	
With head curtains (+ torso bags)	165	331	.498	14 %

TABLE 3-20

MAKE-MODELS THAT ONCE HAD <u>NO</u> SIDE AIR BAGS
AND NOW HAVE <u>STANDARD</u> TORSO BAGS PLUS HEAD PROTECTION,
OR OFFERING A CHOICE BETWEEN NO AIR BAGS AND TORSO + HEAD AIR BAGS
NEARSIDE VS. LONGITUDINAL FATALITIES
(1993-2005 FARS; fatalities adjusted upward for safety belt use;
all cars are 214-certified, equipped with dual frontal air bags, and model year 1994-2003)

Make/models that offered or shifted to torso/head combination bags

	Nearside Fatalities	12:00 or 6:00 Fatalities	Risk Ratio	Nearside Reduction
Without side air bags	702	1,490	.471	
With torso/head combination bags	249	710	.350	26 %

Make/models that offered or shifted to head curtains (plus torso bags)

	Nearside Fatalities	12:00 or 6:00 Fatalities	Risk Ratio	Nearside Reduction
Without side air bags	374	708	.528	
With head curtains (+ torso bags)	184	382	.482	9 %

Make-models that shifted directly from no side air bags to torso bags plus head protection

Approximately half the cars with head air bags to date are make-models that (1) shifted directly from no side air bags to standard torso bags plus head protection, without the intermediate phase of torso bags only; or (2) offered a choice of no air bags or torso plus head air bags in the same year, without offering a choice of torso bags only in that year. These models are well suited for estimating the effect of torso bags plus head protection, because there is no time lag between the cars without side air bags and the cars with both bags. Incidentally, all of the models that went directly from no air bags to standard torso plus head air bags were equipped with combination bags, and many of them are domestic cars (see Table 3-1). But head curtains are included among the models that offered a choice of no air bags or both bags (see Table 3-2).

All three analyses show exceptionally strong fatality reductions for torso plus head air bags. The FARS-GES data show a 38 percent reduction of fatality risk for torso plus head air bags, relative to no bags. The FARS analysis limited to the models with standard torso plus head air bags shows a 40 percent reduction of fatality risk relative to no air bags. When these FARS data are supplemented with the models offering a choice of torso plus head air bags or no bags, the observed effectiveness is 29 percent.

These data strongly support the conclusion that the combination of torso plus head air bags saves lives in nearside impacts, but, of course, they shed no light on the relative contributions of the torso bag and head protection.

Make-models that shifted from torso bags only to torso bags plus head protection

The other half of the cars with head air bags up to model year 2003 are in make-models that (1) shifted from standard torso bags only to standard torso bags plus head protection (and, in most cases, had no side air bags before they had the torso bags); or (2) offered a choice of torso bags or torso plus head air bags in the same year (2003 Honda Accord is the only model offering those choices, and it was also available without any side air bags). These models are, at least in theory, well suited for isolating the benefit of head protection – i.e., estimating the incremental effect of torso bags plus head protection over the effect of torso bags alone. They include most of the head curtains installed to date.

Unfortunately, there are not enough data here to accurately isolate the effect of head protection. For example, the FARS-GES analysis in Table 3-21 shows that these models had a 26 percent lower fatality rate with torso bags plus head protection than with no bags at all. That is generally consistent with all the preceding analyses of torso bags plus head protection versus no bags. But these same models had an equally low fatality rate during the intermediate phase with torso bags only, also 26 percent lower than with no bags. The observed incremental effect of head protection is nil.

TABLE 3-21

MAKE-MODELS THAT SHIFTED FROM TORSO BAGS ONLY
TO TORSO BAGS PLUS HEAD PROTECTION
FATALITIES PER 1,000 NEARSIDE FRONT-SEAT OCCUPANTS IN SIDE IMPACTS
(1993-2005 FARS and GES data; fatalities adjusted upward for safety belt use;
all cars are 214-certified, equipped with dual frontal air bags, and model year 1994-2003)

	Nearside Fatalities	Nearside Occupants	Fatality Rate	Fatality Reduction
Without side air bags	347	94,066	3.69	
With torso bags only	183	67,529	2.72	26 %
With torso bags + head protection	190	70,011	2.72	26 %

Similarly, the FARS analysis limited to models shifting from standard torso bags only to standard torso bags plus head protection (Table 3-22) or supplemented with models that offered these as options at the same time (Table 3-23), show 13 and 14 percent fatality reductions, respectively, for torso bags plus head protection but slightly larger reductions for just the torso bags (16 and 17 percent). The observed incremental benefit of head protection in both tables is -4 percent.

103

TABLE 3-22

MAKE-MODELS THAT SHIFTED FROM TORSO BAGS ONLY
TO TORSO BAGS PLUS HEAD PROTECTION
NEARSIDE VS. LONGITUDINAL FATALITIES
(1993-2005 FARS; fatalities adjusted upward for safety belt use;
all cars are 214-certified, equipped with dual frontal air bags, and model year 1994-2003)

	Nearside Fatalities	12:00 or 6:00 Fatalities	Risk Ratio	Nearside Reduction
Without side air bags	388	707	.549	
With torso bags only	223	486	.460	16 %
With torso bags + head protection	337	476	.478	13 %

TABLE 3-23

MAKE-MODELS SHIFTING FROM TORSO BAGS TO TORSO + HEAD AIR BAGS,
OR OFFERING CHOICES INCLUDING TORSO BAGS AND TORSO + HEAD AIR BAGS
NEARSIDE VS. LONGITUDINAL FATALITIES
(1993-2005 FARS; fatalities adjusted upward for safety belt use;
all cars are 214-certified, equipped with dual frontal air bags, and model year 1994-2003)

	Nearside Fatalities	12:00 or 6:00 Fatalities	Risk Ratio	Nearside Reduction
Without side air bags	395	722	.547	
With torso bags only	230	508	.452	17 %
With torso bags + head protection	230	489	.470	14 %

These analyses, by themselves, with currently available data, cannot be considered meaningful indicators of the effect of head protection. We would hardly conclude that head air bags are relatively ineffective and torso bags account for most of the combined effect. Too many of the earlier analyses, based on substantially larger N, showed higher estimates for torso bags plus head protection. (Also, we have no explanation, other than sampling error, why the effect of torso plus head air bags is lower in these make-models than in the ones that shifted directly from no side air bags to torso plus head air bags.) Nevertheless, these results ought to be a valuable deterrent from jumping to the opposite conclusion [that head protection accounts for most of the benefit, and that torso bags are unnecessary in vehicles equipped with head curtains].

3.5   Fatality reduction in farside impacts of passenger cars

Side air bags have the potential to save lives in farside impacts of passenger cars only if the following four questions are all answered "yes" for at least some crashes:[152]

1. Do side air bags deploy in farside impacts? For example, if a car is occupied only by a driver, and the right-front seat is unoccupied, will the side air bag(s) on the right deploy in a right-side impact?

2. Do occupants make contact with the far side of their car or with the intruding parts of the striking vehicle after a farside impact? For example, will the driver contact any surfaces or components on the right side of the car?

3. Specifically, are these surfaces or components in areas where the air bag deploys?

4. Will the air bag still be fully or at least partially inflated by the time the farside occupant contacts it?

The first question elicits an unequivocal "yes" within the timeframe of this report. As of model year 2007, NHTSA is unaware of any system that, for example, suppresses deployment of side air bags on the right when the right-front seat is unoccupied.

The second question likewise draws an unequivocal "yes." Table 3-24, based on calendar year 1995-2005 data from the Crashworthiness Data System (CDS) of the National Automotive Sampling System (NASS) shows the unweighted distribution of individual life-threatening injuries – levels 4-6 on the Abbreviated Injury Scale (AIS) – to drivers and right-front passengers in farside impacts of passenger cars without side air bags.[153] Table 3-24 shows distributions for

- Unrestrained and belted occupants – and within each of those groups, separately for:
  - Drivers riding alone.
  - All right-front passengers; plus any drivers who were riding with someone else in the front seat.

Surfaces and components located on or beyond the far side of a car are important sources of life-threatening injuries to farside occupants, even belted occupants. Safety belts and/or the presence of other front-seat occupants decrease but hardly eliminate reaching the far side. The wine-colored "subtotal: farside" row of Table 3-24 shows that 74 percent of the life-threatening injuries of unbelted, unaccompanied drivers are due to farside sources such as surfaces and pillars on the far side, direct contact with the striking vehicle and ejection through the farside window or doorway. That is almost as high as in nearside crashes, where 83 percent of life-threatening injuries are from nearside sources.

---

[152] Dainius Dalmotas, in his peer review, asked that this report discuss the rationale whereby side air bags could benefit farside occupants before it presented analyses of farside crash data.

[153] See also Digges, K. and Dalmotas, D, "Injuries to Restrained Occupants in Far-Side Crashes," Paper No. 351, *Proceedings 17th International Technical Conference on the Enhanced Safety of Vehicles*, NHTSA Report No. DOT HS 809 220, Washington, 2001.

TABLE 3-24

PERCENT OF AIS 4-6 INJURIES BY INJURY SOURCE AND BODY REGION FARSIDE IMPACTS, FRONT-SEAT OCCUPANTS OF PASSENGER CARS
(1995-2005 CDS data; cars not equipped with side air bags)

Injury Sources	Body Regions	Unbelted Occupants			Belted Occupants		
		All	Sitting Alone	Not Alone	All	Sitting Alone	Not Alone
Farside, window frame & sill	Head & neck	1	2	0	1	2	0
Far roof rail	Head & neck	2	3	1	2	2	2
Exterior of striking/own vehicle	Head & neck	2	3	1	2	1	3
Farside pillars	Head & neck	11	12	9	5	7	3
Ejected – farside window		9	11	5	0	0	0
Ejected – farside door		2	2	2	1	1	0
Farside interior surface [of door]	Head & neck	15	18	10	20	29	5
Farside interior surface [of door]	Torso & limbs	15	19	9	7	9	4
Farside door armrests/hardware		2	3	0	2	2	1
Farside pillars	Torso & limbs	1	1	0	2	2	0
SUBTOTAL: FARSIDE		60	74	37	42	55	18
Other occupants		2	-	5	7	-	20
Safety belts		-	-	-	4	5	2
Frontal components		13	12	14	13	8	20
Roof (excluding roof rails)		7	4	13	15	15	13
Nearside components & ejection		5	2	9	7	6	9
Interior hardware, other ejections, floor, non-contact		13	8	22	12	11	18
		100	100	100	100	100	100

Table 3-24 shows the percentage decreases to 37 when another person sits between the unstrained occupant and the far side; although this other person is the direct source of only 5 percent of the injuries, there is extensive deflection away from the far side and into the roof, back to the near side and elsewhere. Even for belted but unaccompanied drivers, a remarkable 55 percent of the life-threatening injuries involve contact on the far side. Crash tests and simulations show why: when the impact is close to perpendicular and on the right side, the driver's upper torso readily slips out of the shoulder harness while the pelvis is secured by the lap belt. The driver tips over and, typically, the head contacts the far side, but at a lower spot than the level of the head when the driver is still sitting upright. Indeed, Table 3-24 confirms that the majority of the injuries are head contacts with the interior surface of the door (the orange "29"). That contrasts with nearside impacts, where occupants have no room to tip over and head injuries derive from roof rails, upper pillars or contact with the striking vehicle.

The combination of belt use and the presence of another person reduce the proportion of injuries from farside sources to 18 percent, but even that is not a negligible share.

That leads into the third question: how many of the injuries were in locations that would have been shielded by a deployed side air bag, if the car had been equipped with one? We are not asserting that the air bags will prevent every injury in the areas that they cover, rather that they will not prevent injuries in the areas they do not cover.

- Head curtains cover a large area from front to back, but only as low as the base of the windows. They will stand between the occupant and points around or outside the windows, as well as the upper A- and B-pillars: the red and most of the blue rows in Table 3-24. These rows alone account for 33 percent of all life-threatening injuries for unbelted, lone drivers, but smaller percentages for the other groups. Head curtains are less likely to have an effect for head contacts with the interior surface of the door (the orange row) unless some of them include higher areas or trajectory, and are unlikely to affect the lower contact points (green rows). Inflatable tubular structures have the same front-to-back coverage and perhaps slightly less vertical coverage than head curtains.

- Torso and combination bags deploy in a much narrower zone, front to back. They are designed primarily for nearside impacts, where the occupant is adjacent and likely to contact the bag even if the impact is oblique. But a farside occupant's trajectory can miss the bag. Forward areas such as the A-pillar are certainly not covered. The torso bag covers those parts of the door where it deploys; the combination bag, also the area above that.

The final question is whether the air bag still has any gas pressure by the time a farside occupant has arrived from across the car and made contact. Here, too, head curtains (including inflatable tubular structures) differ from torso and combination bags. Head curtains stay inflated longer because they are not vented; they gradually release gas through the fabric. NHTSA's experience from laboratory and crash testing with the early designs of head curtains is that they inflate in 30-40 milliseconds (msec) and then stay inflated to some extent for at least 150-200 msec and possibly longer, a total cycle of 180-240 msec. (A new generation of head curtains designed to protect in rollovers as well as side impacts stays inflated much longer; however, the crash data analyzed in this report do not include any passenger cars equipped with such curtains.) The agency believes torso bags are fully inflated 30 msec after initial impact and then typically deflate in about 30-40 msec, a total cycle of 60-70 msec. The short cycle helps them accomplish their primary mission, protecting nearside occupants. A combination bag may take 10 msec longer to deflate, for a total cycle of 70-80 msec.

How do these inflation cycles compare to occupant trajectories in actual crashes? A benchmark crash test in Australia involved a Moving Deformable Barrier (MDB) hitting the right side of a large, stationary passenger car, at a 90 degree angle, at 40 mph.[154] The belted driver dummy's head contacted the right door at 150 msec: way too late for torso or combination bags to remain inflated. But the crash test, even though the 40 mph impact speed may sound high by nearside standards, is far less severe than most fatal farside impacts in the United States. Since the Australian MDB weighed 2,178 lbs and the car weighed 3,740 pounds, the Delta V for the car was only 14.7 mph. By contrast, the occupant fatality cases in farside crashes on the 1989-2005 CDS file have a median Delta V of 28 mph for unrestrained front-seat occupants and 29 mph for belted occupants. Because farside occupants are at much lower risk than nearside occupants, the impacts that do kill farside occupants are often extremely severe.

In 90-degree impacts, the lateral distance from the driver's head to the interior surface of the right door divided by the Delta V yields a good first approximation of the time lapse from the beginning of the impact to the contact with the farside air bags.[155] For example, in the crash test, given a distance of 39 inches and a Delta V of 14.7 mph,

$$3.25 \text{ feet} / 21.56 \text{ feet-per-second} = 151 \text{ msec,}$$

close to the 150 msec actually observed on the test. Time-to-contact can be similarly approximated on the farside fatality cases on 1989-2005 CDS with known Delta V. Those cases also specify the curb weight of the struck vehicle. The distance from the driver's head to the farside door was measured at 40 inches on a 3,900-pound passenger car and 33 inches on a 2,400-pound car; let us assume the distance has a linear relationship with curb weight. With that assumption, the percentiles of the time lapse are:

Percentile	Unbelted	Belted
10th	46 msec	41 msec
25th	53	52
50th	68	65
75th	93	86
90th	125	110

---

[154] Bostrom, O., Judd, R., Fildes, B., Morris, A., Sparke, L. and Smith, S, "A Cost Effective Far Side Crash Simulation," *International Journal of Crashworthiness*, Vol. 8, No. 3, 2003, pp. 307-313.
[155] During the intrusion phase of the impact, the farside interior approaches at a speed higher than Delta V; however, that is offset by (1) an initial phase, where the bullet vehicle has contacted the door exterior but the interior surface is not yet moving laterally, and (2) any tendency of the occupant to move laterally – e.g., due to lap belts or friction with the seat.

If these cars had been equipped with head curtains, given an inflation-deflation cycle of 180-220 msec, almost all of them would still have had some gas pressure when the occupant reached the far side – and the majority would still have been close to full pressure. Given cycles of 60-70 msec for torso bags and 70-80 msec for combination bags, approximately half of these bags would still have some positive pressure in them. Relatively few would be close to full pressure (and many of those would be in extremely severe crashes with questionable survivability).

Thus, head curtains (including inflatable tubular structures) have potential for substantial fatality reduction, especially for unrestrained, unaccompanied drivers, to the extent that they cover a wide area front-to-back and in most cases remain inflated until the driver reaches the far side. The effect may be limited because a relatively large proportion of the head injuries in farside impacts are due to contacts below the area shielded by the curtain, especially for belted occupants. Torso bags and combination bags are handicapped by their limited front-to-back coverage and rapid decompression (neither of which is a big problem in nearside impacts). Nevertheless, the above calculations suggest they would still have some gas pressure in many potentially fatal impacts, and could be effective if the occupant's trajectory is toward them. Furthermore, they are the only types that can shield contacts below the window.

In other words, the statistical analyses of crash data must address head curtains (including inflatable tubular structures) and combination bags separately. That will diminish the statistical precision of estimates. Already, the N of farside fatalities is only half of nearside fatalities (see Figure 1-3). Separating the head curtains and the combination bags effectively halves N again. Although head curtains have the best chance of statistically significant results, we must also analyze torso and combination bags. The preceding discussion suggests they are potentially effective, even if just to a limited extent. In addition to the basic analyses for all farside occupants, there should be separate analyses for unbelted and belted occupants, riding alone in the front seat or riding with others, because Table 3-24 showed substantial differences in the distributions of injury sources.

The three analysis methods of the preceding sections will now be applied to farside rather than nearside data: FARS-GES analyses of farside fatality rates per 1,000 occupants in the "core" make-models of Table 3-1 that received side air bags as standard equipment, or in subsets of those make-models; FARS analyses of farside relative to longitudinal crashes in those make-models; and FARS analyses extended to include also models in Table 3-2 with optional side air bags.

FARS-GES analyses Table 3-25 computes fatality rates for drivers and right-front passengers who are farside occupants in side impact crashes – i.e., drivers in right-side impacts and right-front passengers in left-side impacts – for the full "core" group of make-models in Table 3-1. As in preceding sections, the numerator of the fatality rates is the weighted N of FARS cases, with belted cases given a weight greater than 1 to adjust for the effect of belt use; the denominator is weighted GES cases. Without side air bags, the fatality rate is 2.03 per 1,000 exposed occupants, less than half the risk in nearside crashes (4.22 according to Table 3-6). With torso bags only, the fatality rate is 1.88, a non-significant 8 percent reduction.[156] With torso/head combination

---

[156] By the method discussed with Table 3-6, log r = -.0793, standard error = .1646, t = - .48, df = 9.

109

bags, the fatality rate is likewise 1.88, a non-significant 8 percent reduction.[157] But with head curtains (or inflatable tubular structures) plus torso bags the fatality rate drops to 1.41. That is a statistically significant 31 percent reduction from the rate without side air bags. The 90 percent confidence bounds for fatality reduction are 7 to 48 percent.[158]

TABLE 3-25

MAKE-MODELS WITH <u>STANDARD</u> SIDE AIR BAGS
FATALITIES PER 1,000 FARSIDE FRONT-SEAT OCCUPANTS IN SIDE IMPACTS
(1993-2005 FARS and GES data; fatalities adjusted upward for safety belt use;
all cars are 214-certified, equipped with dual frontal air bags, and model year 1994-2003;
make-models with the same type of side impact protection in every year are excluded)

	Farside Fatalities	Farside Occupants	Fatality Rate	Fatality Reduction
Without side air bags	563	276,707	2.03	
With torso bags only	250	133,224	1.88	8 %
With torso/head combination bags	86	45,677	1.88	8 %
With head curtains + torso bags	74	52,855	1.41	31 %

More accurate effectiveness estimates can be obtained by comparing the fatality rate with each respective type of side air bag to the fatality rate in cars of just the same make-models when they did not yet have any side air bags. For example, as in Table 3-12, base fatality rates only on the subset of the "core" make-models that shifted from no side air bags to standard torso bags.[159] As in Table 3-18, compare the fatality rate for make-models with standard combination bags to the rate in these same make-models when they had no side air bags; do likewise for make-models with standard head curtains plus torso bags.[160]

Table 3-26 presents the three estimates. The make-models that shifted from no side air bags to torso bags only had farside fatality rates of 1.88 without side air bags and 1.82 with torso bags, a non-significant 3 percent reduction. A somewhat different group of make-models that shifted from no side air bags to combination bags (either directly or through an intermediate phase of torso bags only) had fatality rates of 2.19 without side air bags and 1.90 with combination bags, a non-significant 14 percent reduction.[161] But the make-models that shifted from no side air bags to head curtains (or inflatable tubular structures) plus torso bags had fatality rates of 2.11 without

---

[157] By the method discussed with Table 3-6, log r = -.0814, standard error = .2766, t = - .29, df = 9.
[158] By the method discussed with Table 3-6, log r = -.3671, standard error = .1591, t = -2.31, df = 9.
[159] And exclude the torso bag cases for make-models that never were produced without side air bags.
[160] Again, exclude make-models that were never produced without side air bags.
[161] By the method discussed with Table 3-6, log r = -.1456, standard error = .2931, t = - .50, df = 9.

side air bags and 1.37 with the curtains.[162] That is a statistically significant 35 percent reduction. The 90 percent confidence bounds for fatality reduction are 7 to 54 percent.[163]

TABLE 3-26

MAKE-MODELS THAT ONCE HAD <u>NO</u> SIDE AIR BAGS
AND NOW HAVE <u>STANDARD</u> SIDE AIR BAGS
FATALITIES PER 1,000 FARSIDE FRONT-SEAT OCCUPANTS IN SIDE IMPACTS
(1993-2005 FARS and GES data; fatalities adjusted upward for safety belt use;
all cars are 214-certified, equipped with dual frontal air bags, and model year 1994-2003)

	Farside Fatalities	Farside Occupants	Fatality Rate	Fatality Reduction
Make/models that shifted from no side air bags to torso bags only				
Without side air bags	389	206,629	1.88	
With torso bags only	238	130,783	1.82	3 %
Make/models that once had no side air bags and later had torso/head combination bags				
Without side air bags	214	97,474	2.19	
With torso/head combination bags	85	44,668	1.90	14 %
Make/models that once had no side air bags and later had head curtains plus torso bags				
Without side air bags	147	69,811	2.11	
With head curtains + torso bags	63	46,172	1.37	35 %

FARS analysis of standard side air bags The effectiveness of side air bags is estimated from FARS data alone by calculating the change in farside fatalities relative to a control group of purely longitudinal impacts (12:00 or 6:00).

- A crash involvement is "farside" if either the initial impact, or the principal impact, or both are on the far side: 2, 3 or 4:00 for drivers and for 8, 9 or 10:00 right-front passengers.

- It is "longitudinal" if the initial and/or the principal impact are at 12:00 or 6:00 while neither impact is a side impact or adjacent to a side impact (same as the nearside analyses).[164]

---

[162] When the analyses in Tables 3-25 and 3-26 are performed separately for head curtains and for inflatable tubular structures, each one shows a large fatality reduction relative to comparison cars without side air bags. The N of cases with inflatable tubular structures is too small for a statistically meaningful comparison of the effectiveness of head curtains and inflatable tubular structures.

[163] By the method discussed with Table 3-6, log r = -.4287, standard error = .1957, t = -2.19, df = 9.

The analysis is based on 1993-2005 FARS data on the core make-models with standard side air bags. As in Table 3-26, the risk ratio of farside to longitudinal fatalities with each respective type of side air bag is compared to the corresponding ratio in cars of just the same make-models when they did not yet have any side air bags. (Similar analyses for nearside occupants may be found in Tables 3-13 and 3-19.) Fatality cases of belted occupants are weighted higher than 1 to estimate the number of fatalities that would have occurred if none of the occupants had been protected by safety belts.[165]

Table 3-27 computes the ratios of farside to purely longitudinal fatalities.

TABLE 3-27

MAKE-MODELS THAT ONCE HAD <u>NO</u> SIDE AIR BAGS
AND NOW HAVE <u>STANDARD</u> SIDE AIR BAGS
FARSIDE VS. LONGITUDINAL FATALITIES
(1993-2005 FARS; fatalities adjusted upward for safety belt use;
all cars are 214-certified, equipped with dual frontal air bags, and model year 1994-2003)

	Farside Fatalities	12:00 or 6:00 Fatalities	Risk Ratio	Farside Reduction
Make/models that shifted from no side air bags to torso bags only				
Without side air bags	452	1,668	.271	
With torso bags only	275	1,193	.230	15 %
Make/models that once had no side air bags and later had torso/head combination bags				
Without side air bags	250	957	.262	
With torso/head combination bags	93	400	.233	11 %
Make/models that once had no side air bags and later had head curtains plus torso bags				
Without side air bags	181	503	.359	
With head curtains + torso bags	74	333	.221	39 %

The make-models that shifted from no side air bags to torso bags had risk ratios of .271 without side air bags and .230 with torso bags, a non-significant 15 percent reduction.[166] The make-

---

[164] For both of these definitions, if one of the impact areas is unknown or not reported, rely on the area that is reported. "A side impact" includes 2-4:00 and 8-10:00 (regardless of where the occupant is sitting) and "adjacent to a side impact" includes 1:00, 5:00, 7:00 and 11:00. All these are excluded, regardless of the occupant's seat position, because they could deploy side air bags.
[165] Kahane (2004), pp. 173-182 and 316-317. Pp. 316-317 estimate belt effectiveness for various types of crashes, based on Kahane (2000), p. 30.
[166] By the method discussed with Table 3-7, log r = -.1623, standard error = .1259, t = -1.29, df = 9.

models that shifted from no side air bags to combination bags experienced a non-significant 11 percent fatality reduction.[167] With head curtains (or inflatable tubular structures) plus torso bags, the fatality reduction is a statistically significant 39 percent. The 90 percent confidence bounds are 12 to 57 percent.[168]

FARS analysis of standard plus optional side air bags Table 3-28 supplements the preceding data with the make-models in Table 3-2 that had optional, but VIN-decodable side air bags in certain years. (Similar analyses for nearside occupants may be found in Tables 3-14 and 3-20.) The first section of Table 3-28 is limited to models that shifted from no side air bags to torso bags and/or offered a choice of no side air bags or torso bags. The fatality reduction for torso bags is a non-significant 3 percent.

TABLE 3-28

MAKE-MODELS THAT ONCE HAD <u>NO</u> SIDE AIR BAGS
AND NOW HAVE <u>STANDARD</u> SIDE AIR BAGS,
OR OFFERING A CHOICE BETWEEN NO AIR BAGS AND SIDE AIR BAGS
FARSIDE VS. LONGITUDINAL FATALITIES
(1993-2005 FARS; fatalities adjusted upward for safety belt use;
all cars are 214-certified, equipped with dual frontal air bags, and model year 1994-2003)

	Farside Fatalities	12:00 or 6:00 Fatalities	Risk Ratio	Farside Reduction
Make/models that offered or shifted to torso bags only				
Without side air bags	647	2,621	.247	
With torso bags only	351	1,471	.239	3 %
Make/models that offered or shifted to torso/head combination bags				
Without side air bags	387	1,490	.260	
With torso/head combination bags	142	710	.200	23 %
Make/models that offered or shifted to head curtains plus torso bags				
Without side air bags	232	708	.327	
With head curtains + torso bags	86	382	.226	31 %

---

[167] By the method discussed with Table 3-7, log r = -.1168, standard error = .2504, t = - .47, df = 9.

[168] By the method discussed with Table 3-7, log r = -.4866, standard error = .1919, t = -2.54, df = 9. Note that the risk ratio with head curtains is not exceptionally low; rather, the ratio in the same make-models without head curtains is high. These models include (1) small cars that may be especially vulnerable in side impact and (2) cars with historically low fatality rates, known to attract safe drivers, who tend to have especially low frontal crash rates. Table 3-27 illustrates the need to compare cars with the various side air bags to cars of the same make-models without the bags; an analysis pooling all the cars without side air bags would have masked the effect of head curtains.

For the make-models that shifted from no side air bags to combination bags or offered them as options, Table 3-28 shows a 23 percent fatality reduction for the combination bags. The reduction is statistically significant, but this will be the only analysis showing a significant effect for combination bags. The 90 percent confidence bounds are 8 to 35 percent.[169] With head curtains (or inflatable tubular structures) plus torso bags, the reduction is a statistically significant 31 percent. Confidence bounds are 9 to 47 percent.[170]

Separate analyses for unrestrained and belted occupants NHTSA believes the belt use of fatally injured occupants is accurately reported on FARS.[171] Table 3-28 is limited to fatality cases. The data may be subdivided and analyzed separately for unbelted and belted occupants.[172] They may be further subdivided by front-seat occupancy: drivers riding alone vs. front-seat occupants who are accompanied by somebody else who sits between them and the far side. When all the cases in a table are unbelted, or all are belted, it is unnecessary to adjust the fatality counts for belt use, as in Table 3-6 – 3-28. Tables can be based on simple, unweighted fatality counts and statistical significance can be established by a chi-square test.

Table 3-29 computes the fatality reduction for head curtains (or inflatable tubular structures) plus torso bags, based on subsets of the data in Table 3-28. For unrestrained drivers and right-front passengers, the risk ratio is .429 without side air bags. It is .180 in the same make-models when they are equipped with head curtains plus torso bags. That is a 58 percent fatality reduction. It is statistically significant at the two-sided .05 level, as evidenced by a chi-square of 10.32. Fatality reductions are a statistically significant 49 percent for unbelted drivers riding alone in the front seat, 68 percent for unbelted front-seat occupants accompanied by another person sitting between them and the far side, and 46 percent for belted, unaccompanied drivers. In other words, the FARS data show significant fatality reductions for all three groups where Table 3-24, based on CDS data, showed substantial proportions of the life-threatening injuries deriving from surfaces or components located on or beyond the far side of the car.[173] (However, each of those three 2x2 tables within Table 3-29 are based on relatively small N – namely, about 25 expected farside fatalities with the head curtains plus torso bags.) Only for belted occupants accompanied by other people in the front seat, where Table 3-24 showed a much smaller proportion of life-threatening head impacts with farside surfaces, does Table 3-30 fail to show a significant or even a positive fatality reduction.

Table 3-30 presents some corresponding analyses for head/torso combination bags. The observed fatality reductions are 20 percent for unbelted occupants and 22 percent for belted occupants. However, neither of them is statistically significant. When the data are further subdivided by front-seat occupancy (not shown in Table 3-30), each result is likewise positive but non-significant.

---

[169] By the method discussed with Table 3-7, log r = -.2584, standard error = .0966, t = -2.67, df = 9.

[170] By the method discussed with Table 3-7, log r = -.3674, standard error = .1499, t = -2.45, df = 9.

[171] Kahane (2000), p. 10; Partyka, S.C., *Lives Saved by Seat Belts from 1983 through 1987*, NHTSA Report No. DOT HS 807 324, Washington, 1988.

[172] "Belted" occupants include REST_USE = 1, 2, 3, 4, 8, 13 and 14 (child safety seats are included); "unbelted" occupants have REST_USE = 0; other occupants (unknown restraint use) are excluded from the analyses.

[173] Furthermore, for these three groups combined, the fatality reduction for head curtains plus torso bags is substantial in small as well as large cars; it is substantial regardless whether the farside impact is by another car, an LTV, a heavy truck or with a fixed object. Only ¼ of the farside fatalities are in crashes where there is a subsequent rollover; rollovers are not what is "driving" the results.

TABLE 3-29

FARSIDE FATALITY REDUCTION BY HEAD CURTAINS + TORSO BAGS
BY RESTRAINT USE AND FRONT-SEAT OCCUPANCY
(1993-2005 FARS; make-models that once had <u>no</u> side air bags and now have <u>standard</u> head
curtains + torso bags, or offering a choice between no air bags and head curtains + torso bags;
farside vs. longitudinal fatalities; actual fatality counts;
all cars are 214-certified, equipped with dual frontal air bags, and model year 1994-2003)

Head Curtains Plus Torso Bags?	Farside Fatalities	12:00 or 6:00 Fatalities	Risk Ratio	Farside Reduction	Chi-Square
ALL UNBELTED OCCUPANTS					
No	85	198	.429		
Yes	20	111	.180	58 %	10.32**
Unbelted, unaccompanied drivers					
No	51	127	.402		
Yes	12	59	.203	49 %	3.71*
Unbelted occupants with another person sitting between them and the far side					
No	34	71	.479		
Yes	8	52	.154	68 %	7.30**
ALL BELTED OCCUPANTS					
No	79	223	.354		
Yes	36	116	.305	14 %	.42
Belted, unaccompanied drivers					
No	48	105	.457		
Yes	16	65	.246	46 %	3.60*
Belted occupants with another person sitting between them and the far side					
No	31	118	.263		
Yes	20	53	.377	- 43 %	1.20

*Statistically significant at the one-sided .05 level (chi-square $\geq$ 2.71).
**Statistically significant at the two-sided .05 level (chi-square $\geq$ 3.84).

TABLE 3-30

FARSIDE FATALITY REDUCTION BY COMBINATION BAGS, BY RESTRAINT USE
(1993-2005 FARS; make-models that once had <u>no</u> side air bags and now have <u>standard</u>
head/torso combination bags, or offering a choice between no air bags and combination bags;
farside vs. longitudinal fatalities; actual fatality counts;
all cars are 214-certified, equipped with dual frontal air bags, and model year 1994-2003)

Head/Torso Combination Bags?	Farside Fatalities	12:00 or 6:00 Fatalities	Risk Ratio	Farside Reduction	Chi-Square
ALL UNBELTED OCCUPANTS					
No	117	413	.283		
Yes	47	208	.226	20 %	1.38
ALL BELTED OCCUPANTS					
No	150	486	.309		
Yes	54	223	.242	22 %	1.86

*Statistically significant at the one-sided .05 level (chi-square $\geq$ 2.71).
**Statistically significant at the two-sided .05 level (chi-square $\geq$ 3.84).

Discussion of findings The CDS analyses of life-threatening injury sources in cars without side air bags, and of the likely time lapse until an occupant contacts the far side provided strong evidence that head curtains (including inflatable tubular structures) have the potential to reduce fatality risk substantially in farside impacts, except for belted occupants when another person sits between them and the far side. They also suggested that combination bags or even torso bags alone could save lives, but to a more limited and less quantifiable extent.

The FARS analyses of the actual crash experience with side air bags directionally confirmed all of these expectations. Every analysis of head curtains (including inflatable tubular structures) plus torso bags showed a statistically significant fatality reduction overall, and also for the subsets of unbelted occupants and of unaccompanied belted occupants.[174] The analyses of combination bags and torso bags showed smaller fatality reductions, but they were not statistically significant except in one analysis of combination bags (Table 3-27).

The only dissonance is that the observed fatality reductions for head curtains seem to be even larger than the potential reductions suggested by Table 3-24. For example, in Table 3-24, up to 33 percent of the life-threatening injuries of unbelted, unaccompanied drivers (2nd column of

---

[174] Furthermore, the significant reductions in farside fatality rates are seen only upon the introduction of head curtains, and not as part of a wider trend. Make-models that did not receive side air bags had nearly constant farside fatality rates throughout MY 1994-2003 (analysis similar to Table 3-9). Make-models that did not receive head curtains had about the same farside fatality rates before and after FMVSS 201 compliance.

numbers) were attributed to farside sources in the upper areas that would be shielded by head curtains (the red and blue rows). But Table 3-29 estimates a 49 percent fatality reduction, suggesting we saved all of these and yet others. In fact, Tables 3-26 – 3-28 show point estimates ranging from 31 to 39 percent for everybody, including belted occupants, including people not sitting alone in the front seat. Two possible explanations:

- The observed fatality reductions derive not just from the head curtains but also in part from the torso bags in these cars. Table 3-24 showed a remarkably high proportion of head impacts and also torso impacts below the window area, and if even a fraction of them are targeted at the bag and arrive before it is deflated, it would appreciably contribute to overall effectiveness.

- The statistically significant point estimates make a strong case for **some** effectiveness but might not accurately indicate its extent. That is not unusual with analyses based on limited data. The confidence bounds for the fatality reduction by head curtains plus torso bags, 7 to 54 percent in Table 3-26, 12 to 57 percent in Table 3-27 and 9 to 47 percent in Table 3-28, include a wide range of values easily compatible with the expectations based on the CDS data.

Even though the specific mechanisms whereby side air bags mitigate injuries in farside impacts have not been extensively demonstrated or quantified by testing, the crash data available at this time make it plausible to conclude that head curtains (including inflatable tubular structures) plus torso bags are beneficial in farside impacts – except for belted occupants when another person sits between them and the far side – but that the effectiveness is somewhat lower than the point estimates in Tables 3-26 – 3-28. Let us assume, for the time being, that the percentage fatality reduction in farside impacts – except for accompanied, belted occupants – is the same as the reduction for all occupants in nearside impacts. That reduction, 24 percent (see Section 3.10) appears to be compatible with the findings from the CDS data as well as the confidence bounds of the FARS estimates. While it is quite possible that combination bags and even torso bags alone also benefit farside occupants, there is not yet enough evidence to uphold a specific, positive effectiveness. NHTSA will update all the effectiveness estimates for farside impacts as more data accumulate.

3.6   Reduction of occupant ejection in side impacts

Occupant ejection through the side window, or through the portal left after the window shatters in a crash, is a major safety problem. Among fatalities that are completely or partially ejected, by far the highest proportion, 56 percent, is ejected via a side window portal. An additional 14 percent are ejected through a side door portal after the door opened in a crash.[175]

Head curtains deploy to cover most of the side window. They are a promising technology for preventing ejection through the side window portal. Head/torso combination bags, although covering a smaller area than curtains, might prevent some ejections. Torso air bags deploying from the seat could conceivably prevent some ejections through open doors. For maximum

---

[175] *National Highway Traffic Safety Administration Evaluation Program Plan, Calendar Years 2004-2007*, NHTSA Report No. DOT HS 809 699, Washington, 2004, p. 30, based on calendar year 1995-2003 NASS CDS data for model year 1990-2003 cars and LTVs.

effectiveness, head curtains should also deploy in rollover crashes, where an exceptionally high proportion of fatalities are ejected. Ford developed a sensor system to deploy "Safety Canopy" head curtains in rollover crashes and began to offer it on selected SUVs in mid-model year 2002.[176] Subsequently, other manufacturers also introduced rollover sensors.

Through model year 2003, however, side air bags in passenger cars were designed to deploy primarily in side impacts. At least that provides a method for evaluating their effectiveness in preventing ejection. Side air bags might prevent fatal ejections in side impacts but should have little effect on fatal ejections in other impacts – frontals, first-event rollovers, rear impacts – because the bags do not deploy. Fatal ejections in other impacts become a control group. The ratio of ejected fatalities in side impacts to ejected fatalities in other impacts could decrease with side air bags.

Table 3-31 compares ejected fatalities in side impacts to other impacts in passenger cars without side air bags, to cars with the various types of side air bags. Complete and partial ejections are included. "Side impacts" are those where the initial and/or principal impact point is 2, 3, 4:00 or 8, 9, 10:00. All other crash involvements, including non-collisions such as first-event rollovers are "other impacts" (except that vehicles with unknown initial and principal impact are excluded). Because the N of fatal ejections is limited, all passenger cars of model years 1997-2003 on calendar year 1996-2005 FARS are included, not just the make-models that were equipped with side air bags. In those model years, all cars were equipped with dual air bags and met the dynamic test requirement of FMVSS 214. Cases have not been weighted to adjust for belt use. With unweighted FARS data, statistical significance can be tested with chi-square.

TABLE 3-31

EJECTED FATALITIES IN SIDE IMPACTS VS. OTHER IMPACTS
ALL MODEL YEAR 1997-2003 PASSENGER CARS
(1996-2005 FARS, drivers and right-front passengers)

| | Ejected Fatalities | | Risk | |
	Side Impacts	Other Impacts	Ratio	Reduction
Without side air bags	1,802	4,336	.416	
With torso bags only	150	333	.450	- 8 %
With torso/head combination bags	64	181	.354	15 %
With head curtains[177]	39	134	.291	30 %

From cars without side air bags, 1,802 drivers and right-front passengers were ejected and fatally injured in side impacts and 4,336 in other impacts: a risk ratio of .416. With torso air bags alone,

---

[176] http://www.ford.com/en/innovation/safety/rolloverProtection.htm.
[177] Most of these cars also have torso bags, but cars with head curtains only are included here.

the risk ratio is .450, a non-significant 8 percent increase.[178] With torso/head combination bags, there is a non-significant 15 percent reduction relative to no side air bags.[179]

But cars equipped with head curtains (or inflatable tubular structures) experienced 39 fatal ejections in side impacts and 134 in other impacts, a risk ratio of .291; the 30 percent reduction relative to cars without side air bags is statistically significant at the one-sided .05 level.[180] Whereas Table 3-31 is still based on fairly limited data, the evidence so far is that head curtains are helpful in reducing ejections.

3.7   Fatality reduction by side air bags in LTVs: early results

Side air bags were introduced later on light trucks and vans (LTVs – comprising pickup trucks, SUVs, minivans and full-size vans up to 10,000 pounds Gross Vehicle Weight Rating) than on passenger cars. Torso bags were introduced in 1998 on the Chevrolet Venture, Oldsmobile Silhouette, Pontiac Trans Sport vans and on Mercedes SUVs. Head protection, in the form of torso/head combination bags, was offered on Ford Explorer and Mercury Mountaineer SUVs in 1999. LTV cases with side air bags on FARS through 2005 are only a fraction of the number for passenger cars.

Table 3-32 analyzes the effect of side air bags in LTVs to date in **nearside** impacts, based on the ratio of nearside to longitudinal fatalities in 1994-2005 FARS. In order to maximize N, any make-model that ever offered side air bags as standard equipment or as a VIN-decodable option is included for model years 1995-2003, even make-models where only a small proportion of the vehicles were equipped with side air bags, and even make-models where they were standard from the beginning. Also included are the predecessors of these make-models back to 1995: LTVs with a different name, but the same manufacturer and addressing basically the same market segment (e.g., Jeep Cherokee and Jeep Liberty). No pickup trucks or full-size vans are included in Table 3-32, because hardly any of the former and none of the latter were equipped with side air bags by 2003. In other words, all the LTVs in Table 3-32 are SUVs or minivans. The analysis is limited to drivers and right-front passengers at seats equipped with frontal air bags. FARS cases have not been weighted to adjust for belt use.

---

[178] Chi-square for the 2x2 table made by the first two rows of Table 3-31 is 0.62.

[179] Chi-square for the 2x2 table made by the first and third rows of Table 3-31 is 1.19.

[180] Chi-square for the 2x2 table made by the first and fourth rows of Table 3-31 is 3.78.

TABLE 3-32

LTV MAKE-MODELS WITH STANDARD OR OPTIONAL SIDE AIR BAGS
NEARSIDE VS. LONGITUDINAL FATALITIES
(1994-2005 FARS; model year 1995-2003;
drivers and right-front passengers at seats equipped with frontal air bags)

	Nearside Fatalities	12:00 or 6:00 Fatalities	Risk Ratio	Nearside Reduction
Without side air bags	1,652	3,897	.424	
With torso bags only	177	484	.366	14 %
With torso bags + head protection	54	173	.312	26 %

LTVs without side air bags experienced 1,652 nearside and 3,897 longitudinal fatalities, a risk ratio of .424. There are just 177 nearside fatality cases with torso bags only and 54 with torso bags plus head protection (much smaller N than the 727 and 448, respectively, for passenger cars in Table 3-8). With torso bags only, the risk ratio is .366. That is a non-significant 14 percent reduction from the risk ratio without side air bags.[181] With torso bags plus head protection, the risk ratio is .312. That is a 26 percent reduction, and it is statistically significant at the one-sided .05 level.[182] Both point estimates are quite close to our best point estimates of fatality reduction in passenger cars (12 and 24 percent, respectively: see Section 3.10). We may assume, until we obtain enough LTV data that would lead us to conclude otherwise, that side air bags are about as effective in nearside impacts for LTVs as for cars.

In **farside** impacts, the N of LTVs specifically equipped with head curtains is insufficient for statistically meaningful separate analyses.[183] Moreover, we cannot simply assume that the effect of head curtains is the same as in cars. The discussion for cars in Section 3.5 about farside injury sources and the time lapse for an occupant to reach the far side may not apply to LTVs, because of their greater mass, width and height. We are unable at this time to assert head curtains are effective in farside impacts of LTVs, let alone claim any specific, positive effectiveness estimate.

3.8   Child passengers and side air bags

Side air bags deploy much less forcefully than frontal air bags, and are not expected to present a significant risk to child passengers. Nevertheless, crash data need to be monitored for cases where a child passenger could have potentially contacted a deploying side air bag – i.e., where child passengers age 0 to 12 were seated in positions equipped with side air bags and involved in a nearside impact. We should check for cases involving low speeds or severity that would not ordinarily result in fatalities, or any reports of injuries that could be related to deploying air bags.

---

[181] Chi-square for the 2x2 table formed by the first two rows of Table 3-32 is 2.55.

[182] Chi-square for the 2x2 table formed by the first and third row of Table 3-32 is 3.75.

[183] Nine farside fatality cases, no observed fatality reduction relative to LTVs without side air bags.

NHTSA's Special Crash Investigations (SCI) Program is a prime source of detailed information about the performance of side (and frontal) air bags in crashes. Specifically, SCI endeavors to find any crash where occupants might have been seriously injured by deploying air bags. As of November 2006, SCI is not aware of a single case in the United States where a child age 12 years or younger (or, for that matter, any other person) was fatally injured by a deploying side air bag in an otherwise survivable, low-to-moderate speed crash. (By contrast, at that time SCI had 174 confirmed cases of child fatalities related to frontal air bags in otherwise survivable crashes.)

FARS, complete through the end of calendar year 2005, contained 18 cases of child passenger fatalities in nearside impacts seated at a position equipped with a side air bag(s), including 8 right-front passengers of cars, 2 right-front passengers of SUVs, 1 right-front passenger of a minivan and 7 rear-outboard passengers of cars.[184] The 7 rear-seat positions were equipped with head curtains only; 4 right-front seats were equipped with torso bags plus head curtains, 6 with torso bags only, and 1 with a combination bag.

None of these 18 cases appear to have been "otherwise survivable low-to-moderate speed crashes" or raise any suspicion about the performance of the side air bags. In all 18, the case vehicle was towed away with disabling damage, according to FARS. Seven of the crashes were fatal to at least one person other than a child seated next to an air bag. Among the other 11 case vehicles, one was struck by a train, one impacted a bridge pier, and one experienced a most-harmful-event rollover subsequent to the side impact; the remaining eight were struck in the side by substantially heavier vehicles.

Likewise, a statistical analysis of the ratio of nearside to frontal fatalities, similar to Table 3-8 but limited to child passengers age 0-12 (and combining all types of side air bags into a single category), does not show a statistically significant difference with and without the air bags.

3.9 Vehicles with head curtains only: early results

Beginning in model year 2001, Saturn, Chrysler Sebring and Dodge Stratus offered head curtains only, without torso bags, as an option for the front seat. As of model year 2004, only a few other passenger cars offer them. Head curtains only are available on several SUV models, including some with high sales. Beginning in model year 2002, Jeep Grand Cherokee and Liberty, Dodge Durango, Ford Explorer and Mercury Mountaineer were available with head curtains, recognizable from the VIN. In rear seats (not analyzed in this report), head curtains without torso bags are the only type of side air bags as of model year 2006. With the availability of 2005 FARS data, an initial statistical analysis is possible.

Table 3-33 analyzes the effect of head curtains only in **nearside** impacts, based on the ratio of nearside to longitudinal fatalities in 1999-2005 FARS, separately for passenger cars and SUVs. In order to maximize N, any make-model that ever offered head curtains only as standard equipment or as a VIN-decodable option is included for model years 1995-2006, even make-

---

[184] Includes all VIN-decodable side air bags through model year 2006.

models where only a small proportion of the vehicles were equipped with head curtains only.[185] The analysis is limited to drivers and right-front passengers at seats equipped with frontal air bags. FARS cases have not been weighted to adjust for belt use.

The upper section of Table 3-33 analyzes passenger cars. N is limited: there are just 21 nearside fatality cases with head curtains only. The risk ratio is .876 without side air bags and .636 with head curtains only. That is a non-significant 27 percent fatality reduction.[186] The lower section tabulates the SUV data. Here, too, the fatality reduction is a non-significant 27 percent.[187]

Can the car and SUV data be combined to produce a single, statistically more meaningful result? Simply pooling the data is inadvisable because the nearside risk ratio is so much lower in SUVs than in cars, and a higher proportion of the SUVs in the table are equipped with head curtains. The GENMOD procedure of SAS/STAT® (logistic regression that allows categorical independent variables) calibrates the proportion of fatalities in nearside impacts as a function of vehicle type (car or SUV) and equipment (head curtains or no side air bags). The regression coefficient for head curtains is -.320, equivalent to a $1 - \exp(-.320) = 27$ percent reduction in fatality risk (appropriately, the same as in the separate analyses of Table 3-33). Because the chi-square for the coefficient is 2.02, this effect is also non-significant.

TABLE 3-33

MAKE-MODELS WITH STANDARD OR OPTIONAL HEAD CURTAINS ONLY
NEARSIDE VS. LONGITUDINAL FATALITIES
(1999-2005 FARS; model year 1995-2006;
drivers and right-front passengers at seats equipped with frontal air bags)

	Nearside Fatalities	12:00 or 6:00 Fatalities	Risk Ratio	Nearside Reduction
PASSENGER CARS				
Without side air bags	183	209	.876	
With head curtains only	21	33	.636	27 %
SUVs				
Without side air bags	78	250	.312	
With head curtains only	12	53	.226	27 %

---

[185] For make-models with optional head curtains, the analysis includes only the model years when they were available as an option; for make-models with standard head curtains, the analysis also includes, for comparison purposes, preceding year(s) when no side air bags were available.
[186] Chi-square for the 2x2 table is 1.16.
[187] Chi-square for the 2x2 table is .87.

The point estimates for head curtains only, although not statistically significant, are close our best estimates for torso bags plus head protection (see next section). We are not yet able to conclude that head curtains alone are effective, let alone establish a meaningful estimate of fatality reduction, but at least these early results do not raise any alarm.

3.10   Best effectiveness estimates

Every statistical analysis of nearside impacts to passenger cars showed fatality reduction by **torso air bags plus head protection**. Three analyses may be singled out because they include a wide range of cars with standard and/or optional torso bags plus head protection, but compare their fatality risk only to cars of the same make-models without any side air bags:

- Table 3-15, based on FARS and GES data, compares the nearside fatality rate per 1,000 occupants with standard torso bags plus head protection to the rate in the same make-models without side air bags. The fatality reduction is a statistically significant 31 percent.

- Table 3-16, based on FARS, compares the ratio of nearside to longitudinal (impact location 12:00 or 6:00) fatalities with standard torso bags plus head protection to the ratio in the same make-models without side air bags. The fatality reduction is a non-significant 20 percent.

- Table 3-17 expands the analysis of Table 3-16 to also include make-models that offer a choice between no bags and torso bags plus head protection. With the additional data, the fatality reduction is a statistically significant 19 percent.

The average of those three key results, **24 percent**, will serve as the best point estimate of the fatality reduction by torso bags plus head protection in nearside impacts to passenger cars.[188] Confidence bounds for that estimate should take into account not only sampling error (the result of limited data) but computational uncertainty (as evidenced by varied results when different analyses are applied to fundamentally the same data). The 90 percent confidence bounds range from **4 to 42 percent**.[189]

Furthermore, the analyses of this chapter generated a total of fifteen point estimates for the fatality reduction by torso bags plus head protection in nearside impacts of passenger cars: Tables 3-6, 3-7, 3-8, 3-8a (in Section 3.2), two regressions in Section 3.2, Tables 3-15, 3-16 and 3-17 (in Section 3.4), three analyses of make-models that shifted directly from no bags to torso plus head air bags (in Section 3.4), and three analyses of cars that had torso bags only before they had torso plus head air bags (Tables 3-21, 3-22 and 3-23). From lowest to highest, these fifteen estimates are:

---

[188] The actual effectiveness estimates (not rounded) are 31.29%, 20.18% and 19.37%.
1 - exp { [log(1 - .3129) + log(1- .2018) + log(1-.1937)] / 3 } = 1 – exp (-.2720) = 23.81 percent
[189] When expressed as log r, the three estimates are -.3753, -.2254 and -.2153 and have standard errors .1418, .1510 and .0926, respectively. The confidence bounds will use the lowest and highest point estimates (computational error) and the tightest standard error (sampling error). The lower bound is 1 – exp (-.2153 + 1.833x.0926) = 4%. The upper bound is 1 – exp (-.3753 - 1.833x.0926) = 42%. (–1.833 is the 5th percentile of a t-distribution with 9 degrees of freedom, the appropriate multiplier for the sampling error of the first estimate.) This approach tries to address Dalmotas' guidelines for the error analysis.

13, 14, 19, 19, 20, 21, 21, 26, 26, 29, 31, 34, 35, 38 and 40 percent

All estimates are positive. The median of the fifteen estimates is 26 percent, a bit higher than 24 percent. But the analyses of control groups at the end of Section 3.2 (Tables 3-9, 3-10 and 3-11) hinted at a slight general tendency toward lower fatality rates in later model years, even without side air bags. They suggested that point estimates for side air bags should be viewed with some caution: when different analyses generate a range of point estimates, the best answer is probably somewhat lower than the median of that range.

There is less certainty about the fatality reduction for **torso bags only**. Three principal analyses are:

- Table 3-12, based on FARS and GES, compares the nearside fatality rate per 1,000 occupants with standard torso bags plus to the rate in the same make-models without side air bags. The fatality reduction is a statistically significant 15 percent.

- Table 3-13, based on FARS, compares the ratio of nearside to longitudinal fatalities with standard torso bags to the ratio in the same make-models without side air bags. The fatality reduction is a statistically significant 17 percent.

- Table 3-14 expands the analysis of Table 3-13 to include make-models that offer a choice between no bags and torso bags. With the additional data, the fatality reduction is a non-significant 5 percent.

The average of those three key results, **12 percent**, will serve as the best point estimate of the fatality reduction by torso bags only in nearside impacts to passenger cars.[190] The 90 percent confidence bounds, calculated as above, range from **–3 to +23 percent**.[191] In other words, considering the variation in the principal estimates, we cannot definitely conclude at this time that torso bags alone reduce fatality risk.

This chapter also generated nine other point estimates for torso air bags; here are all twelve estimates, listed from lowest to highest:

-2, 2, 5, 13, 14, 15, 16, 17, 17, 17, 26 and 26 percent

These estimates are not so consistent. They range from suggesting that torso air bags alone have little or no benefit to saying they account for most of the benefit of torso bags plus head protection. The median of the point estimates is 15.5 percent, and, as above, the best estimate probably should be somewhat below the median.

If torso bags plus head protection reduce fatality risk by 24 percent (23.81 percent, not rounded) and torso bags alone by 12 percent (12.39 percent, not rounded), the incremental fatality

---

[190] The actual effectiveness estimates (not rounded) are 15.06%, 16.79% and 4.87%.
1 - exp { [log(1 - .1506) + log(1- .1679) + log(1-.0487)] / 3 } = 1 – exp (-.1323) = 12.39 percent
[191] When expressed as log r, the three estimates are -.1632, -.1838 and -.0499 and have standard errors .0439, .0875 and .0817, respectively. The confidence bounds will use the lowest and highest point estimates (computational error) and the tightest standard error (sampling error). The lower bound is 1 – exp (-.0499 + 1.833x.0439) = –3%. The upper bound is 1 – exp (-.1838 - 1.833x.0439) = 23%.

reduction for torso plus head air bags over torso bags alone is 13 percent.[192] In other words, torso bags and head protection each make important and, in a sense, more or less equal contributions to saving lives in side impacts.

The effectiveness of **head curtains alone**, without torso bags, is unknown because few vehicles had been equipped that way through 2005, although a number of make-models had recently received that configuration. Early analyses, as shown in Section 3.9 are not unfavorable, although they do not generate statistically significant results. Intuitively, torso bags plus head protection ought to be the safest configuration, because torso bags alone do not provide head protection in nearside impacts, while head curtains alone would not protect the torso.

All the preceding estimates are for nearside impacts, both single- and multi-vehicle, to passenger cars. Section 3.7 suggests that, for the time being, the estimates can be accepted for nearside impacts to LTVs as well as cars.

Section 3.5 showed statistically significant fatality reductions for head curtains (including inflatable tubular structures) plus torso bags in farside impacts to passenger cars – for any unrestrained front-seat occupants and also for belted drivers who rode alone in the front seat. Although the FARS analyses in Section 3.5 generated higher estimates, the distribution of injury sources in CDS data suggests scaling those estimates back somewhat – say, to the fatality reduction for head curtains plus torso bags in nearside impacts (24 percent). At this time, we are unable to estimate a fatality reduction, if any, in farside impacts for LTVs (with any type of side air bags); for torso or combination bags in cars; and for head curtains for belted occupants, when somebody sits between them and the far side.

---

[192] $1 - [(1-.2381) / (1-.1239)]$

CHAPTER 4

LIVES SAVED AND SAVABLE IN 2003 BY SIDE IMPACT PROTECTION

4.0   Summary

In calendar year 2003, 9,107 drivers and passengers were fatalities in cars and LTVs whose initial and/or principal impact was a side impact. If none of these cars and LTVs had been equipped with side air bags, and if all the cars had side impact performance characteristic of the pre-1986 fleet, fatalities would have increased to 9,979. Thus, TTI(d) reduction in cars and side air bags in cars and LTVs saved an estimated 872 lives in 2003. But if all cars and LTVs on the road in 2003 had been equipped with head curtains plus torso bags and if every car had post-1996 side impact performance, there would have been only 7,045 fatalities. In other words, the combination of TTI(d) reduction and head curtains plus torso bags could have saved an estimated 2,934 lives (9,979 – 7,045) if it had been installed in the entire on-road fleet by 2003.

4.1   A model for estimating lives saved and savable by side impact protection

In 2004, NHTSA estimated the *Lives Saved by the Federal Motor Vehicle Safety Standards and Other Vehicle Safety Technologies, 1960-2002* by developing a model that operated on FARS data.[193] That model, however, did not address certain technologies that had begun to appear in production cars and LTVs but whose effectiveness had not yet been evaluated by statistical analyses of crash data – including:

- FMVSS 214 – side impact protection, dynamic test standard for passenger cars

- Torso air bags for cars and LTVs

- Head-protection air bags for cars and LTVs

Using the same general method, we will define a new model that estimates the lives saved by these three technologies, in side impacts, in calendar year 2003.

Each 100 actual fatality cases on FARS represent a theoretically even greater number of fatalities that could have happened if the vehicles had not been equipped with side impact protection. The actual fatality cases in side impacts are the starting point for a model to estimate the lives saved by side impact protection.

For example, FARS might have records of 100 driver fatalities in model year 2003 four-door cars equipped with torso bags plus head-protection air bags, in multivehicle crashes with principal impact point 9:00 (i.e., nearside impacts). In Section 3.10 we estimated that torso bags with head protection reduce fatality risk by 24 percent in nearside impacts. If the cars had not been equipped with side air bags, there would have been not 100 but

---

[193] Kahane, C.J., *Lives Saved by the Federal Motor Vehicle Safety Standards and Other Vehicle Safety Technologies, 1960-2002,* NHTSA Technical Report No. DOT HS 809 833, Washington, 2004, pp. x-xi, 2-3, 11, 173-182 and 281-360.

$$100/(1 - .24) = 132 \text{ would-be fatalities.}$$

In other words, the existence of these 100 actual fatality cases with torso bags plus head protection on FARS implies that the MY 2003 cars were involved in 132 crashes that could have been fatal to their drivers. However, 32 of those could-have-been fatal crashes did not become FARS fatality cases because the air bags saved the driver's life.

Upgraded structures and/or padding in response to or in anticipation of the dynamic test requirement of FMVSS 214 reduced the TTI(d) of 4-door cars from an average of 85 before model year 1986 to an average of 63 after FMVSS 214 certification. Section 2.6 estimates that a 22-unit reduction in TTI(d) corresponds to a 17.3 percent fatality reduction for nearside occupants in multivehicle crashes. If, in addition to removing side air bags, the side structures in those cars had been degraded to pre-1986 levels, the number of would-be fatalities would have increased from 132 to

$$132/(1 - .173) = 159 \text{ would-be fatalities.}$$

The TTI(d) reduction saved $159 - 132 = 27$ lives. The 100 actual fatality cases on FARS imply the existence of 159 could-have-been fatal crash cases. Side impact protection (torso plus head air bags, TTI(d) reduction by structure and padding) saved a combined total of 59 lives: 32 by side air bags and 27 by structure and padding.

In other words, the model begins with the actual FARS fatality cases and inflates them step-by-step, using the effectiveness estimates from Sections 2.6 and 3.10, until side impact protection has been "removed" from the vehicle – until the vehicle has been degraded to a level of safety performance characteristic of before 1986 rather than its actual model year. The safety technologies are "removed" in the reverse chronological order that they were introduced: first remove the side air bags, then remove the padding and downgrade the structure. At each step into the past, the model tallies the lives saved by the latest safety technology – i.e., the additional fatalities that would have occurred if that technology had been removed.[194]

To this point, the model estimates the number of lives **actually** saved in calendar year 2003 by the side impact protection that was already in the vehicles on the road in that year. But the model can also be extended to estimate the number of lives that could **hypothetically** have been saved in calendar year 2003 if every car and LTV had been equipped with head curtains plus torso bags, and if every car on the road, no matter what its model year, had side structures and padding characteristic of post-FMVSS 214 cars.[195]

The latter is **not** an estimate of how many lives will be saved per year, eventually, if all vehicles on the road were fully equipped with side impact protection. Several factors would diminish the number of lives potentially saved per year:

---

[194] The model produces unbiased estimates of the lives saved, without double counting, because the effectiveness estimates in Chapters 2 and 3 are based on analyses of vehicles produced just after vs. just before a technology was introduced. Specifically, all of the analyses of side air bags were limited to cars that were already certified to FMVSS 214, even the cars in these analyses without the side air bags. The effectiveness estimates in Chapter 3 are no more than the incremental effect of side air bags, given that a car already meets FMVSS 214.

[195] This extension of the model was not undertaken in the 2004 report.

- The introduction of other safety technologies, especially Electronic Stability Control in many vehicles should reduce the number of potentially fatal side impacts.[196]

- The long-term shift of the on-road fleet from 2-door to 4-door cars, and from passenger cars to SUVs will reduce the number of potentially fatal side impacts.

NHTSA regulatory analyses predict annual future benefits and assess their uncertainty. This evaluation report does not. The objective here is only to supply some point estimates of absolute numbers to flesh out the estimates of relative effectiveness (i.e., percentage fatality reductions) of the preceding chapters – in the hypothetical situation that all vehicles on the road in 2003 had already been equipped with head curtains plus torso bags and met FMVSS 214.[197]

The analyses of this chapter focus exclusively on side impacts and do not include possible benefits in rollover crashes – not yet evaluated from crash data – of head curtains with rollover sensors. They focus on drivers and right-front passengers and exclude benefits – not yet evaluated from crash data – of side air bags, structures or padding for rear-seat occupants.

To see how the extension of the model works, consider a population of MY 1997, 4-door cars, all certified to meet FMVSS 214, but none equipped with side air bags. Let us say there were 132 driver fatalities in nearside impacts by other vehicles. The first part of the model estimates what happens if TTI(d) had degraded from 63 to 85: fatalities would have increased from 132 to 159. Thus, structures and padding actually saved 27 lives. The second part of the model starts with these fully downgraded cars (now equivalent to pre-MY 1986 cars with unimproved side structures and no side air bags) and works the process in reverse. First, TTI(d) is "restored" from 85 back to 63. Fatalities would decrease by 17.3 percent, from 159 back to 132. Next, head curtains plus torso bags are also introduced in the cars. That would reduce the fatalities by 24 percent, from 132 to 100. Thus, side structures and padding **actually** saved 27 lives (from 159 to 132), but the combination of structures, padding, head curtains and torso bags would **hypothetically** have saved 59 lives (from 159 to 100), relative to the number of would-be fatalities with pre-1986 technology.

This model focuses on the side impact protection evaluated in Chapters 2 and 3. It does not include the effects of side door beams meeting the original static strength requirement of FMVSS 214 that went into effect for passenger cars on January 1, 1973 and for LTVs on September 1, 1993. According to NHTSA's 2004 report, side door beams saved an estimated 489 occupants of passenger cars and 146 occupants of LTVs in calendar year 2002, a total of 635 lives saved. Readers, if they wish, may add 635 to the estimates of this chapter to estimate lives saved by all side impact protection technologies implemented since 1960.[198]

The 2004 report also estimated that "voluntary" TTI(d) reductions in 2-door cars saved 359 lives in calendar year 2002.[199] The estimates of this chapter (803 lives saved in 2003 by all –

---

[196] Dang, J.N., *Preliminary Results Analyzing the Effectiveness of Electronic Stability Control (ESC) Systems*, NHTSA Evaluation Note No. DOT HS 809 790, Washington, 2004.

[197] Dainius Dalmotas and John Jacobus, in their peer reviews of this report, emphasized the distinction between lives hypothetically savable in 2003 and lives potentially savable in future years.

[198] Kahane (2004), pp. 217 and 241. The inclusion of side door beams would have necessitated, essentially, a complete update of the model in the 2004 report.

[199] *Ibid.*, pp. 223-224.

voluntary plus post-FMVSS 214 – TTI(d) reductions before side air bags, including 437, specifically, in 2-door cars) supersede (and contain within them) the estimates in the 2004 report.

4.2   Parameters for the model

Actual fatalities Calendar year 2003 has been selected as the root year for the computations because the FARS file is complete and fixed, and because 2003 is the last model year for most of the statistical analyses of side air bags. The model will be applied to FARS data for the full calendar year 2003.

The model is applied to the 9,107 fatality cases of driver and right-front passengers who were occupants of cars and LTVs involved in side impacts.

- Consistent with the FARS analyses of the effectiveness of side air bags, a "side impact" is a crash-involved vehicle case with:
  - Initial and/or principal impact 2, 3, 4, 8, 9 or 10:00.
  - First harmful event not a rollover, fire, immersion, or fall from a moving vehicle (codes 1-6).
  - Moreover, if any of the impacts are on the occupant's side, the case is a "nearside" impact, even if another impact is on the far side. The remaining cases are "farside" impacts, having at least one side damage, but none on the near side.
- Cars and LTVs are defined by the VIN, if decodable, and by the FARS body type, otherwise. All model years are included.
- Drivers and right-front passengers are defined by the seat-position variable on FARS.

By 2003, missing data on any of these variables had become rare on FARS (less than 3 percent, total); unlike the 2004 report on lives saved by the FMVSS, cases with missing data on impact location, seat position or vehicle type are deleted, not imputed.

Availability of side air bags For model years 1996-2004, the presence and type of side air bags at the occupant's seat position are decoded from the VIN; if the VIN is unknown, they are decoded from the FARS make-model whenever possible. No vehicles were equipped with side air bags before 1996.

In 185 of the 9,107 cases, the availability of side air bags in a specific vehicle could not be determined because:

- Air bags were optional, but their presence could not be identified from the VIN.
- Air bags were optional, and the VIN was unknown.
- The make-model or the model year was unknown.

When the make-model and model year were known, it was possible to assign that vehicle a probability, corresponding to the proportion of sales equipped with torso bags, or with torso and head air bags, or with head curtains only, based on *Ward's Almanacs* percentage of factory-installed side air bags and the information in Tables 3-2 and 3-4. For example, 17 percent of 2001 Honda Civics were equipped with torso bags. Two high-sales make-models with optional side air bags that usually cannot be decoded from the VIN are Honda Civic and Mitsubishi Galant. In yet other cases it was possible to conclude from the model year (even if the make-model was unknown) or the make-model (even if the model year was unknown) that a vehicle could not have been equipped with side air bags. That left only 24 case vehicles without any clue about side air bags.

These 24 cases are conservatively treated on both steps of the model. When computing lives actually saved to date, let us assume these vehicles were not equipped with side air bags. But subsequently, when computing hypothetically savable lives, let us assume they are already equipped with head curtains plus torso bags. The model will not estimate any lives saved to date or hypothetically savable for these vehicles of uncertain status.

Effectiveness of side air bags The best estimates in Section 3.10 are that torso bags plus head protection reduce fatality risk by 24 percent in **nearside** impacts and torso bags alone, by 12 percent. These fatality reductions are averages that apply to right-front passengers as well as drivers, single- and multi-vehicle crashes, LTVs as well as cars. As discussed in Sections 1.4, 3.3 and 3.4, these effectiveness estimates may include some benefits actually due to energy-absorbing materials (other than air bags) installed to meet the FMVSS 201 upgrade of head-impact protection; however, the portion not attributable to air bags is probably minimal because these energy-absorbing materials remained largely unchanged at the time head air bags were installed in the make-models analyzed.

Section 3.10 has no effectiveness estimate for head curtains only. There were 14 side impact fatalities in 2003 at seats equipped with head curtains only. Let us assume a 12 percent fatality reduction in nearside impacts, the same effect as torso bags alone. That is an educated guess, based on the incremental effect of torso plus head air bags over torso bags alone, and because the life-threatening injuries in side impacts include approximately equal numbers of head and torso injuries (Section 1.1).

In **farside** impacts to passenger cars, based on the discussion in Sections 3.5 and 3.10, we are assuming that head curtains plus torso bags reduce fatality risk by 24 percent (same effect as for nearside impacts), but only for unrestrained occupants and for belted drivers who are sitting alone in the front seat. For head curtains only, 12 percent. No fatality reduction is assumed in farside impacts at this time: for any type of side air bag in LTVs, for combination or torso bags in cars, and for belted car occupants when another person sits between them and the far side of the car.

In models such as Honda Civic or Mitsubishi Galant, where the presence of side air bags in a specific vehicle cannot be deduced from the VIN, the effectiveness of side air bags is multiplied by the probability that a vehicle was equipped with them. For example, in 2001 Honda Civic (17 percent equipped with torso bags only), the effectiveness in nearside impacts is .12 x .17 = 2.04 percent.

A similar approach is used for farside fatality cases with unknown belt use, when another person sits between the fatally injured occupant and the far side of the car. In calendar year 2003, when belt use or non-use was reported (1,085 cases), 51.6 percent of farside fatalities (not sitting alone) were restrained. Let us assume, likewise, a belt use rate of 51.6 percent among the 94 cases with unknown restraint use. In these cases, the effectiveness of side air bags is multiplied by the probability that the occupant was unrestrained: 48.4 percent. For example, the fatality reduction by head curtains plus torso bags for farside fatality cases (not sitting alone) with unknown belt use is .24 x .484 = 11.62 percent.

TTI(d) improvements without side air bags Before the first side air bags were introduced in 1996, side impact protection was significantly improved by structures and/or padding in many passenger cars after FMVSS 214 certification, and to some extent even before that. The model year of initial FMVSS 214 certification is known for each make-model of passenger car. It is 1994, 1995, 1996 or 1997. This part of the model applies only to passenger cars. Although LTVs also certified to FMVSS 214, and side structures in some of the smaller LTVs such as compact pickup trucks may have been upgraded, we have no estimate of the extent of TTI(d) improvement, if any, and will not assume any in this analysis (see Sections 1.8 and 2.5).

The injury criterion TTI(d) is a measure of injury risk in side impacts. NHTSA does not have complete information on TTI(d) performance by make-model and model year, especially for the model years before FMVSS 214. Instead, we will use average values of TTI(d), depending on certification status, the number of doors, and the model year, based on the analyses in Section 1.5.

Two-door cars of model years 1981-1985, at least the models tested by NHTSA and others, had an average TTI(d) of 114, well above the 90 subsequently allowed by FMVSS 214. Model years 1981-1985 were chosen as the baseline level of TTI(d) performance because: (1) NHTSA does not have any earlier TTI(d) measurements on production cars, and (2) They are at the beginning of the rulemaking process to upgrade FMVSS 214. Performance began to improve after 1985 as manufacturers began to "design to TTI(d)" and, by 1993-1996 TTI(d) averaged 95 even for the models that were not yet certified to FMVSS 214. As shown in Section 1.5, the average TTI(d) of 214-certified 2-door cars without side air bags was 69, and changed little from year to year. Table 4-1 shows the average values of TTI(d), by model year, that will be assumed in the model. Rather than using the actual year-to-year averages from 1986 to 1992 in Table 1-2 (which, in any individual year are based on relatively few test vehicles), let us assume a gradual improvement from 1986 to 1993, accelerating in the later years, as shown in Table 4-1.

Similarly, 4-door cars had an average baseline TTI(d) of 85 in 1981-1985, almost exactly what FMVSS 214 subsequently allowed for these cars. Performance began to improve after 1985 and had reached an average of 71 by 1993-1996, substantially better than what the standard would allow. The average TTI(d) of 214-certified 4-door cars without side air bags is 63, and has changed little from year to year. Table 4-1 shows the average values of TTI(d) that will be assumed in the model; it assumes a gradual improvement from 1986 to 1993, accelerating in the later years.

In all, TTI(d) in 2-door cars was improved by an average of 45 units without side air bags, from 114 in the pre-1986 baseline to 69 after certification. It improved by an average of 22 units, from 85 to 63, in 4-door cars.

TABLE 4-1

AVERAGE TTI(d) BY MODEL YEAR
(parameters used in the "lives saved" model)

Model Year(s)	2-Door Cars	4-Door Cars
1985 and earlier	114	85
1986	112	84
1987	110	83
1988	108	82
1989	106	81
1990	104	80
1991	102	78
1992	99	75
1993-1996, not 214 certified	95	71
214-certified (no side air bags)	69	63

Effect of TTI(d) improvements on fatalities Section 2.6 calibrated a 0.863 percent[200] fatality reduction in nearside impacts with other vehicles per one-unit improvement in TTI(d). It did not show a statistically significant fatality reduction for single-vehicle nearside impacts or any farside impacts – i.e., little evidence to support any specific positive estimate of fatality reduction there.

The preceding fatality reductions were calibrated from the crash experience of 15 make-models that substantially improved TTI(d) without side air bags, and where the TTI(d) was actually measured in tests before and after the improvement. These models' TTI(d) averaged 85 before and 62 after the improvement. The model years of the cars in the analysis ranged from 1991 to 2002; 80 percent of the cars had TTI(d) less than 90; 70 percent were 4-door cars; and 58 percent of the cars were FMVSS 214-certified. In other words, these data are especially appropriate for calibrating the effect of TTI(d) improvements in the below-90 range, for a mix of 2- and 4-door cars in the years shortly before and after the FMVSS 214 phase-in.

But the analyses of this report are not so appropriate for calibrating the effects of TTI(d) improvement in a higher range, typical of 2-door cars of the 1980's and early 1990's. NHTSA's Phase 1 evaluation of side impact protection calibrated the effect of TTI(d) improvement from crash data exclusively for 2-door make-models of model years 1981-1993; 16 of the 17 make-models in that analysis (94 percent) had TTI(d) over 90; none were FMVSS 214-certified. There, fatality risk in all types of side impacts (average of nearside and farside, single-vehicle and multi-vehicle) decreased by 0.927 percent per one-unit improvement in TTI(d). In other

---

[200] $1 - \exp(-.00863)$.

133

words, the effect is apparently somewhat stronger in the higher ranges of TTI(d), and applicable in a wider range of side impacts.[201]

For the "lives saved" model we will use the 0.927 percent effect for improvements in the average TTI(d) when that average is above 90 and the 0.863 percent effect when the average is 90 or less (and the latter includes all averages for 4-door cars, as may be seen in Table 4-1). More realistically, of course, the effect would not change abruptly at 90 or any specific other point but would transition smoothly in some range well above and below 90; however, we have no basis for establishing a transition rate or a range. Since we are selecting a specific "boundary," 90 is an intuitively good choice, even though the datasets in the two analyses slightly overlap on both sides of 90, because it is the FMVSS 214 requirement for 2-door cars. Essentially, most 2-door cars until just before FMVSS 214 had TTI(d) over 90, whereas most 4-door cars even before the standard, and of course all 214-certified cars had scores below 90.

The percentage fatality reduction is computed by finding the number of units of TTI(d) improvement and applying the percentage reduction per unit that number of times, similar to computing net present value (compound interest at a negative rate of interest). For example, 4-door cars had a baseline TTI(d) of 85, improving to 63 after 214 certification, a difference of 22 units. The entire improvement is within the "below 90" range. In multivehicle nearside impacts, the fatality reduction for a 214-certified car relative to a baseline car is

$$1 - \exp(-.00863 \times 22) = 17.3 \text{ percent}$$

In single-vehicle nearside impacts and in farside impacts, we are not assuming any fatality reduction. Similar formulas are used to compute the effect of voluntary TTI(d) improvements for 4-door cars of model years 1986-1996 that were not 214-certified, except "22" is replaced by the difference of the baseline TTI(d), 85, and the TTI(d) for non-certified cars of that model year as shown in Table 4-1.

Two-door cars had a baseline TTI(d) of 114, improving to 69 after 214 certification, a difference of 45 units. However, the first 24 units of the improvement are in the "above 90" range and only the last 21 are in the "below 90" range. In multivehicle nearside impacts, the fatality reduction for a 214-certified car relative to a baseline car is

$$1 - [(1 - .00927)^{24} \times \exp(-.00863 \times 21)] = 33.29 \text{ percent}$$

In single-vehicle nearside impacts and in farside impacts, the estimated reduction is

$$1 - (1 - .00927)^{24} = 20.03 \text{ percent}$$

Voluntary TTI(d) improvements for 2-door cars of model years 1986-1996 that were not 214-certified were all in the "above 90" range. One fatality reduction is estimated for all types of side impacts, and it is

---

[201] Kahane, C.J., *Evaluation of FMVSS 214 - Side Impact Protection: Dynamic Performance Requirement; Phase 1: Correlation of TTI(d) with Fatality Risk in Actual Side Impact Collisions of Model Year 1981-1993 Passenger Cars*, NHTSA Technical Report No. DOT HS 809 004, Washington, 1999, p. 84.

$$1 - (1 - .00927)^{[\,114 - TTI(MY)\,]}$$

As discussed in Section 2.5, no fatality reduction is assumed for compact pickup trucks or any other LTVs.

4.3   Results

The model processes each of the 9,107 actual side-impact fatality cases on the 2003 FARS. Each case is assigned an original weight factor of 1. The model first "removes" the side air bags from the vehicle (if there were any in it) and then "degrades" the TTI(d) performance to 1985 baseline levels (if the vehicle was a post-1985 car), inflating the case weight factor as each technology is removed and fatality risk increases. Subsequently, the model "upgrades" the TTI(d) performance to post-FMVSS 214 levels (if the vehicle was a car), and then "installs" head curtains plus torso bags, deflating the case weight factor as each potential technology is added, enhancing survival.

For example, the case weight of a nearside fatality in a 214-certified 4-door car without side air bags in a multivehicle crash would start at 1 and inflate to 1/(1-.173) = 1.21 after TTI(d) is degraded to the baseline 1985 level. Subsequently, restoring TTI(d) to post-FMVSS 214 levels would at first deflate the case weight back to 1, and the addition of head curtains plus torso bags would deflate it to .76. Thus, the fatality case contributes evidence of .21 lives actually saved by TTI(d) improvement without side air bags and .24 lives that could have been saved if the car had been equipped with head curtains plus torso bags, totaling up to .45 lives actually saved and/or hypothetically savable by side impact protection. Benefits are summed and tabulated by safety technology, by vehicle type, and in the case of TTI(d) improvements, by whether or not the car was 214-certified.

Lives actually saved in 2003   Table 4-2 estimates that side air bags and TTI(d) improvements saved a total of 872 lives in calendar year 2003.

TABLE 4-2[202]

LIVES SAVED BY SIDE IMPACT PROTECTION IN CALENDAR YEAR 2003
(drivers and right-front passengers of cars and LTVs;
fatality reduction for the actual on-road fleet in 2003
relative to a fleet with MY 1985 baseline side impact protection)

Side air bags in cars and LTVs		
Head curtains plus torso bags	17	
Torso/head combination bags	24	
Torso bags only	27	
Head curtains only	1	
Subtotal: side air bags		69
TTI(d) improvement by structures and padding in cars		
In FMVSS 214-certified cars	569	
Voluntary improvements in 1986-1996 non-certified cars	234	
Subtotal: TTI(d) improvements		803
TOTAL: LIVES SAVED IN 2003		872

Even though many of the latest cars and LTVs have side air bags, the proportion of all vehicles on the road equipped with side air bags was still quite small in 2003. In that year, side air bags saved 69 drivers and right-front passengers, including 17 at seats equipped with head curtains plus torso bags, 24 with combination bags and 27 with torso bags alone.

By contrast, the majority of cars on the road in 2003 were FMVSS 214-certified, and an even larger proportion at least incorporated some level of voluntary improvement over the 1985 baseline level of side impact protection. These TTI(d) improvements saved an estimated 803 lives in 2003, including 569 in the 214-certified cars and 234 in pre-standard cars with voluntary improvements.

Table 4-3 categorizes the benefits by vehicle type. Side air bags saved 59 lives in passenger cars and 10 in LTVs during 2003. That disparity reflects the earlier installation of side air bags in cars as well as the greater risk to car occupants in side impacts.

Even though 74 percent of the cars on the road in 2003 were 4-door models, TTI(d) improvements saved 437 lives in 2-door cars and only 367 in 4-door cars. That reflects, of course, the much worse baseline performance of 2-door cars and their substantially greater TTI(d) improvement from model year 1985 to 1997.

---

[202] All estimates in Table 4-2 through 4-6 have been rounded to the nearest integer, although the model can and does estimate in fractional amounts. Subtotals may not add up exactly in these tables because of the rounding. In this table, for example, the estimates of lives saved by four types of side air bags were actually 16.72, 23.86, 26.65 and 1.41 (rounded to 17, 24, 37 and 1 in the table). They add up to 68.64 (rounded to 69).

TABLE 4-3

LIVES SAVED BY SIDE IMPACT PROTECTION IN 2003 – BY VEHICLE TYPE
(drivers and right-front passengers of cars and LTVs;
fatality reduction for the actual on-road fleet in 2003
relative to a fleet with MY 1985 baseline side impact protection)

	2-Door Cars	4-Door Cars	LTVs
Head curtains plus torso bags	2	15	0
Torso/head combination bags	2	17	5
Torso bags only	4	18	5
Head curtains only	0	1	0
Subtotal: side air bags	8	51	10
TTI(d) improvements in 214-certified cars	293	277	
Voluntary, pre-standard TTI(d) improvements	144	90	
Subtotal: TTI(d) improvements	437	367	
TOTAL: LIVES SAVED IN 2003	445	417	10

Lives hypothetically savable in 2003 Table 4-4 shows that side impact protection could have saved an estimated 2,934 lives in calendar year 2003 if every car and LTV had been equipped with head curtains plus torso bags and if every car had been equipped with side structures and padding characteristic of FMVSS 214-certified vehicles. In other words, the number of side impact fatalities that would have happened in 2003 if every car and LTV on the road had been so equipped is 2,934 fewer than the number that would have occurred if no vehicles at all had side air bags and if all cars still had TTI(d) performance characteristic of 1985 and earlier model years. The hypothetical benefits include 1,791 lives savable by head curtains plus torso bags in cars and LTVs and 1,143 lives savable by TTI(d) improvements in cars.

TABLE 4-4

LIVES HYPOTHETICALLY SAVABLE BY SIDE IMPACT PROTECTION IN 2003
(drivers and right-front passengers of cars and LTVs;
fatality reduction for a fleet of 214-certified vehicles with head curtains plus torso bags
relative to a fleet of MY 1985 baseline vehicles)

Head curtains plus torso bags	1,791
TTI(d) improvement by structures and padding in cars	1,143
TOTAL: LIVES HYPOTHETICALLY SAVABLE IN 2003	2,934

NHTSA's 1990 Regulatory Impact Analysis for FMVSS 214 predicted how many lives would be decreasing TTI(d) to various levels: specifically that at least 512 lives would be saved per year if TTI(d) improved from its baseline levels in cars of the mid-1980's to 90 or better in all 2-door cars and 85 or better in all 4-door cars.[203] Those are the just-passing levels specified in FMVSS 214. The 1,143 lives saved, as estimated in Table 4-4, are more than double NHTSA's prediction. The principal reason for such a favorable result, of course, is that manufacturers actually improved TTI(d) well beyond the just-passing levels, to an average of 69 in 214-certified 2-door cars without side air bags, and 63 in 4-door cars.

Lives actually saved, as shown in Table 4-2 are a subset of the hypothetically savable lives in Table 4-4. Lives actually saved by side air bags (69) are only a fraction of hypothetical savings with head curtains plus torso bags (1,791), because most vehicles on the road in 2003 did not have any side air bags and, among those that did, many had only a torso bag, a combination bag or a head curtain. By contrast, TTI(d) improvements had already achieved (803) the majority of their potential (1,143) because most of the cars on the road by 2003 were FMVSS 214-certified or at least incorporated some voluntary improvements.

In summary, there were 9,107 side impact fatalities on FARS in 2003. If none of the vehicles had been equipped with side air bags or TTI(d) improvements, fatalities would have increased to 9,979 (9,107 fatalities plus 872 actual lives saved). But if every vehicle on the road in 2003 had already been equipped with head curtains plus torso bags and if every car had been 214-certified, there would have been only 7,045 fatalities (9,979 minus 2,934 lives hypothetically savable).

Table 4-5 shows that head curtains plus torso bags could have saved 1,443 lives in passenger cars and 348 in LTVs during calendar year 2003, again reflecting the higher vulnerability of passenger cars in side impact crashes. The majority of the hypothetical benefits of TTI(d) improvements would have been in 2-door cars.

TABLE 4-5

LIVES HYPOTHETICALLY SAVABLE IN 2003 – BY VEHICLE TYPE
(drivers and right-front passengers of cars and LTVs;
fatality reduction for a fleet of 214-certified vehicles with head curtains plus torso bags
relative to a fleet of MY 1985 baseline vehicles)

	2-Door Cars	4-Door Cars	LTVs
Head curtains plus torso bags	427	1,016	348
TTI(d) improvement by structures and padding	649	494	___
LIVES HYPOTHETICALLY SAVABLE IN 2003	1,076	1,510	348

---

[203] *Final Regulatory Impact Analysis - New Requirements for Passenger Cars to Meet a Dynamic Side Impact Test FMVSS 214*, NHTSA Publication No. DOT HS 807 641, Washington, 1990, p. IV-62.

Percent of baseline fatalities savable by side impact protection The model can also calculate the combined, overall effectiveness of head curtains plus torso bags and TTI(d) improvements based on structure and padding. It is the number of hypothetically savable lives divided by the baseline fatalities in a fleet with neither type of side impact protection – i.e., actual fatalities plus lives saved in 2003.

Table 4-6 carries out the calculations. The 24 fatality cases where nothing is known about whether the vehicle was equipped with side air bags, due to unknown make-model and/or model year are excluded in Table 4-6.[204] There were 2,112 fatalities in 2-door cars. Adding the 445 lives actually saved in 2003 generates an estimate of 2,557 baseline fatalities if none of these cars had side air bags and average TTI(d) were degraded to the baseline, 1981-1985 average level of 114. If all of these baseline cars then received structure and/or padding that improved their TTI(d) to an average of 69 and also gained head curtains plus torso bags, 1,076 lives could have been saved in 2003. The combined effect of the TTI(d) improvements and the head curtains plus torso bags is a 1,076/2,557 = 42 percent fatality reduction in side impacts.

TABLE 4-6

PERCENT OF BASELINE FATALITIES SAVABLE BY SIDE IMPACT PROTECTION
(drivers and right-front passengers of cars and LTVs)

	2-Door Cars	4-Door Cars	LTVs
Actual side-impact fatalities in 2003[205]	2,112	4,660	2,311
Lives saved by side impact protection in 2003	445	416	10
Fatalities in baseline fleet w/o side impact protection	2,557	5,076	2,321
Lives hypothetically savable by side impact protection	1,076	1,510	348
Percent of baseline fatalities savable	42%	30%	15%

Similarly, in 4-door cars, the combined effect of reducing average TTI(d) from the baseline average of 85 to the post-standard average of 63 and installing head curtains plus torso bags is a 30 percent fatality reduction. In LTVs, the effect of head curtains plus torso bags is a 15 percent fatality reduction, averaged over all side impacts: we estimated a 24 percent fatality reduction in nearside impacts, but do not at this time claim a benefit in farside impacts. Even though side impacts continue to be a vulnerable point of passenger cars and a cause of many fatalities, a combination of improved side structures, padding, head curtains and torso bags greatly reduces fatality risk to occupants.

---

[204] Because the earlier, conservative assumption that no lives are savable by side air bags in these vehicles (and even by TTI(d) improvements if the model year is unknown) would lead to an understatement of potential effectiveness.
[205] Excluding cases where nothing is known about whether the vehicle was equipped with side air bags, due to unknown make-model and/or model year.

REFERENCES

Bostrom, O., Judd, R., Fildes, B., Morris, A., Sparke, L. and Smith, S. "A Cost Effective Far Side Crash Simulation," *International Journal of Crashworthiness*, Vol. 8, No. 3, 2003, pp. 307-313.

Braver, E.R. and Kyrychenko, S.Y. *Efficacy of Side Airbags in Reducing Driver Deaths in Driver-Side Collisions.* Insurance Institute for Highway Safety, Arlington, VA, 2003.

Buying a Safer Car. NHTSA Publication No. DOT HS 809 546, Annual publication, 2003-2005.

Buying a Safer Car 2000. NHTSA Publication No. DOT HS 809 046, Washington, 2000.

Buying a Safer Car 2001. NHTSA Publication No. DOT HS 809 152, Washington, 2000.

Buying a Safer Car 2002. NHTSA Publication No. DOT HS 809 409, Washington, 2002.

Creating a Market for Safety – 10 Years of Euro NCAP. European New Car Assessment Programme, Brussels, 2005, accessible from www.euroncap.com .

Dang, J.N. *Preliminary Results Analyzing the Effectiveness of Electronic Stability Control (ESC) Systems.* NHTSA Evaluation Note No. DOT HS 809 790, Washington, 2004.

Digges, K. and Dalmotas, D. "Injuries to Restrained Occupants in Far-Side Crashes," Paper No. 351, *Proceedings 17th International Technical Conference on the Enhanced Safety of Vehicles.* NHTSA Report No. DOT HS 809 220, Washington, 2001, accessible from www-nrd.nhtsa.dot.gov/departments/nrd-01/esv/esv.htm .

Evans, L. *Traffic Safety and the Driver.* Van Nostrand Reinhold, New York, 1991.

Federal Register Notices:

> 33 (October 5, 1968): 14971, ANPRM announcing the intention to regulate side door strength.

> 35 (October 30, 1970): 16801, Final Rule establishing FMVSS 214 (side door strength, passenger cars).

> 55 (October 30, 1990): 45752, Final Rule upgrading FMVSS 214, adding a dynamic side impact test for passenger cars.

> 56 (June 14, 1991): 27427; Final Rule extending FMVSS 214 side door strength requirement to LTVs.

> 58 (October 4, 1993): 51735, Executive Order 12866 – Regulatory Planning and Review.

60 (July 28, 1995): 38749, Final Rule extending the dynamic side impact test of FMVSS 214 to LTVs up to 6,000 pounds GVWR.

60 (August 18, 1995): 43031, Final Rule upgrading FMVSS 201.

63 (August 4, 1998): 41451, Final Rule reducing the FMVSS 201 test speed from 15 mph to 12 mph on target areas where a head air bag is stored.

68 (May 17, 2004): 27990, NPRM to amend FMVSS 214, adding a 20 mph side impact with a pole.

Final Regulatory Impact Analysis - New Requirements for Passenger Cars to Meet a Dynamic Side Impact Test FMVSS 214. NHTSA Publication No. DOT HS 807 641, Washington, 1990.

Ford Safety Canopy System (http://www.ford.com/en/innovation/safety/rolloverProtection.htm).

Gabler, H.C. and Hollowell, W.T. *The Aggressivity of Light Trucks and Vans in Traffic Crashes.* Paper No. 980908, Society of Automotive Engineers, Warrendale, PA, 1998.

_____. "NHTSA's Vehicle Aggressivity and Compatibility Research Program," Paper No. 98-S3-O-01, *Proceedings of the 16th International Technical Conference on the Enhanced Safety of Vehicles.* Report No. DOT HS 808 759, Washington, 1998, accessible from www-nrd.nhtsa.dot.gov/departments/nrd-01/esv/esv.htm .

Government Performance and Results Act of 1993. Public Law 103-62, August 3, 1993.

Hedeen, C.E. and Campbell, D.D. *Side Impact Structures.* Paper No. 690003, Society of Automotive Engineers, New York, 1969.

Insurance Institute for Highway Safety, Vehicle Ratings. www.iihs.org/ratings .

Joksch, H. *Vehicle Design versus Aggressivity.* NHTSA Technical Report No. DOT HS 809 184, Washington, 2000.

Kahane, C.J. *Evaluation of FMVSS 214 - Side Impact Protection: Dynamic Performance Requirement; Phase 1: Correlation of TTI(d) with Fatality Risk in Actual Side Impact Collisions of Model Year 1981-1993 Passenger Cars.* NHTSA Technical Report No. DOT HS 809 004, Washington, 1999.

_____. *An Evaluation of Side Structure Improvements in Response to Federal Motor Vehicle Safety Standard 214.* NHTSA Technical Report No. DOT HS 806 314, Washington, 1982.

_____. *Fatality Reduction by Air Bags: Analyses of Accident Data through Early 1996.* NHTSA Technical Report No. DOT HS 808 470, Washington, 1996.

_____. *Fatality Reduction by Safety Belts for Front-Seat Occupants of Cars and Light Trucks.* NHTSA Technical Report No. DOT HS 809 199, Washington, 2000.

_____. *Lives Saved by the Federal Motor Vehicle Safety Standards and Other Vehicle Safety Technologies, 1960-2002*. NHTSA Technical Report No. DOT HS 809 833, Washington, 2004.

_____. *Vehicle Weight, Fatality Risk and Crash Compatibility of Model Year 1991-99 Passenger Cars and Light Trucks*. NHTSA Technical Report No. DOT HS 809 662, Washington, 2003.

Kahane, C.J. and Hertz, E. *The Long-Term Effectiveness of Center High Mounted Stop Lamps in Passenger Cars and Light Trucks*. NHTSA Technical Report No. DOT HS 808 696, Washington, 1998.

Ludtke, N.F., Osen, W., Gladstone, R. and Lieberman, W. *Perform Cost and Weight Analysis, Head Protection Air Bag Systems, FMVSS 201*. NHTSA Technical Report No. DOT HS 809 842, Washington, 2004.

_____. *Perform Cost and Weight Analysis, Non Air Bag Head Protection Systems, FMVSS 201*. NHTSA Technical Report No. DOT HS 809 810, Washington, 2003.

McCartt, A.T. and Kyrychenko, S.Y. *Efficacy of Side Airbags in Reducing Driver Deaths in Driver-Side Car and SUV Collisions*. Insurance Institute for Highway Safety, Arlington, VA, 2006.

McNeill, A., Haberl, J., Holzner, M., Schoeneburg, R., Strutz, T. and Tautenhahn, U. "Current Worldwide Side Impact Activities – Divergence versus Harmonisation and the Possible Effect on Future Car Design," Paper No. 05-0077, *Proceedings 19th International Technical Conference on the Enhanced Safety of Vehicles*. NHTSA Report No. DOT HS 809 825, Washington, 2005, accessible from www-nrd.nhtsa.dot.gov/departments/nrd-01/esv/19th/esv19.htm .

Morgan, C. *The Effectiveness of Retroreflective Tape on Heavy Trailers*. NHTSA Technical Report No. DOT HS 809 222, Washington, 2001.

National Highway Traffic Safety Administration Evaluation Program Plan, Calendar Years 2004-2007. NHTSA Report No. DOT HS 809 699, Washington, 2004.

NHTSA Hails Safety Features in Model Year 1994 Passenger Cars and Light Trucks and Vans. Press Release No. NHTSA 38-93, U. S. Department of Transportation, Office of the Assistant Secretary for Public Affairs, Washington, 1993.

NHTSA Plan for Achieving Harmonization of the U.S. and European Side Impact Standards, Report to Congress, April 1997. NHTSA Docket No. NHTSA-1998-3935-1, 1998.

Partyka, S.C. *Lives Saved by Seat Belts from 1983 through 1987*. NHTSA Report No. DOT HS 807 324, Washington, 1988.

Passenger Vehicle Identification Manual. Annual Publication, National Insurance Crime Bureau, Palos Hills, IL.

Preliminary Economic Assessment, FMVSS 214, Amending Side Impact Dynamic Test Adding Oblique Pole Test. NHTSA Docket No. NHTSA-2004-17694-1, 2004.

Preliminary Economic Assessment, NPRM for Light Trucks, Buses and Multipurpose Passenger Vehicle, Dynamic Side Impact Protection, FMVSS No. 214. NHTSA Docket No. 88-06-N23-001, 1994.

SAS/STAT® User's Guide, Vol. 1, Version 6, 4th Ed. SAS Institute, Cary, NC, 1990.

Side Impact Conference. NHTSA Report No. DOT HS 805 614, Washington, 1980.

Status Report, Vol. 38, August 26, 2003, Insurance Institute for Highway Safety, Arlington, VA.

Tarbet, M.J. *Cost and Weight Added by the Federal Motor Vehicle Safety Standards for Model Years 1968-2001 in Passenger Cars and Light Trucks.* NHTSA Technical Report No. DOT HS 809 834, Washington, 2004.

Traffic Safety Facts 2004, NHTSA Report No. DOT HS 809 919, Washington, 2005.

Walz, M.C. *Evaluation of FMVSS 214 Side Impact Protection for Light Trucks: Crush Resistance Requirements for Side Doors.* NHTSA Technical Report No. DOT HS 809 719, Washington, 2004.

APPENDIX A

MAKE-MODEL GROUPS FOR EVALUATION OF FMVSS 214

The make-model groups and the larger categories defined here are used in Chapter 2 for evaluating the effect of major TTI(d) reductions that did not involve side air bags.

A "make-model group" may consist of a single make-model over several model years (during which time it may be redesigned), or of two or more make-models that, in any given model year, are very similar vehicles (such as Ford Taurus and Mercury Sable). The make-model groups are largely carried over from Chapter 8 and Appendix B of the Phase 1 evaluation report for FMVSS 214.[206]

The make-model groups are assigned to four larger categories based on the history of their TTI(d) test results and on their vehicle modifications, as documented by their manufacturers in response to NHTSA information requests (IR):

1. Make-model groups that substantially improved TTI(d) (namely, reduced it by 9 or more units) upon FMVSS 214 certification <u>or</u> at some later time, <u>without</u> side air bags. If there is information about the vehicle's modifications, it usually shows major new structure. When we assign make-model groups to this category we must also specify <u>when</u> TTI(d) decreased.

2. Make-model groups with no TTI(d) change or limited change, as evidenced by: actual TTI(d) history showing little or no change, or IR unequivocally stating that the car certified to FMVSS 214 without any changes from the preceding model year..

3. Make-model groups where NHTSA does not know the amount of TTI(d) change, if any, because the agency has no test data prior to certification; however, the IR perhaps suggests they received major new structure upon FMVSS 214 certification

4. Other make-model groups with unknown TTI(d) change upon FMVSS 214 certification; the IR, if there is one, does not indicate they received major new structure.

Within each of the four categories, the make-model groups are listed in the order of the 5-digit FARS codes for the predominant make-model in the group.[207] (The order of the make-model codes in FARS is Chrysler, Ford, General Motors, and the foreign-based nameplates in more-or-less alphabetical order.)

---

[206] Kahane, C.J., *Evaluation of FMVSS 214 - Side Impact Protection: Dynamic Performance Requirement; Phase 1: Correlation of TTI(d) with Fatality Risk in Actual Side Impact Collisions of Model Year 1981-1993 Passenger Cars*, NHTSA Technical Report No. DOT HS 809 004, Washington, 1999, pp. 139-155 and 185-238. Some groups from the Phase 1 report are omitted in this report because they were discontinued by 1996, never certified to FMVSS 214, or never tested for TTI(d). Others are omitted because we have only a single TTI(d) test result – i.e., no history of change over time – and/or because we lack information on how/if the vehicles were modified to meet FMVSS 214. Three groups not included in the Phase 1 report because their sales were relatively low in the early-to-mid 1990's have been added here because their test histories indicate a substantial decrease in TTI(d), without air bags, upon certification or upon a subsequent remodeling.

[207] The group numbers in the Phase 1 report have been omitted here because they are not needed for the analysis.

The "TTI(d) test history" tables for these make-models describe the specific cars tested by NHTSA. Side air bags are optional for some models in certain years. The tables indicate whether NHTSA tested a car equipped with the optional air bags, an unequipped car, or even one of each (e.g., 2001 Honda Civic 2-door).

Category 1: Make-model groups that substantially improved TTI(d) upon FMVSS 214 certification or at some later time, without side air bags

Dodge Intrepid/Chrysler Concorde/Eagle Vision 4-door

TTI(d) test history:

Model Year	Adjusted TTI(d)[208]	N of Tests	Side Air Bags?	Car Body Platform (Wheelbase and Other Shared Features)[209]
1993	78.8	1		6025 CHRYSLER LH CARS, 93-
1994	64.5	1		6025 CHRYSLER LH CARS, 93-
1997	54.8	1		6025 CHRYSLER LH CARS, 93-
1998	50.7	1		6025 CHRYSLER LH CARS, 93-
1999	49.8	1		6025 CHRYSLER LH CARS, 93-

Initial FMVSS 214 certification: 1994

Vehicle modifications (described in response to NHTSA's Information Request): 1994 Intrepid had modified beams, C pillar, roof bow, reinforced sills to strengthen the B-pillar, cross-member (no picture, but sounds substantial from the long list of changes); no mention of padding. 1998 Intrepid has additional reinforcement of the B-pillar and a redesigned beam.

Conclusions: TTI(d) was 79 in the original 1993 Intrepid. It appears substantial structural modifications were made upon self-certification in 1994, improving TTI(d) to 65. It is also possible, but not quite so evident, that a second round of modifications in 1997 or 1998 further improved TTI(d) to about 50 without air bags.

Assignment: Category 1, substantial TTI(d) improvement without air bags upon original FMVSS 214 certification in 1994.

Ford Mustang 2-door

TTI(d) test history:

1988	110.0	1		12027 FORD MUSTANG 100, 79-93
1996	57.8	1		12038 FORD MUSTANG 101.3, 94-
1997	67.3	1		12038 FORD MUSTANG 101.3, 94-
2001	65.1	2		12038 FORD MUSTANG 101.3, 94-

---

[208] TTI(d) adjusted = TTI(d) observed * (33.54/actual test speed)^2 as explained in Section 1.5.
[209] These 5-digit codes are used in NHTSA evaluations to indicate cars that share the same wheelbase, body platform and other features. They supplement the 5-digit FARS make-model codes.

Initial FMVSS 214 certification: 1996

Vehicle modifications (IR): 1996 Mustang received moderate structural revisions in the door beams, door inner belts and a cross-car member, energy absorbing foam in the door trim panels.

Conclusions: Ford Mustang was redesigned in 1994. It is most likely that the high TTI(d) in the 1988 model carried over until 1993, and that a very substantial improvement was already "built into" the 1994 redesign. The changes in the 1996 model described above sound pretty small for a TTI(d) improvement of 50 units; thus, it is plausible that a large decrease took place with the major redesign of 1994, and a smaller decrease in 1996. Quite probably, Ford had no need to certify the 1994 Mustang (because they had other models exceeding the required 10 percent of sales that they could certify without any modifications), and certified it in 1996 after some additional improvements.

Assignment: Category 1, substantial TTI(d) improvement without air bags upon the 1994 redesign (and not upon the subsequent FMVSS 214 certification in 1996).

Ford Taurus/Mercury Sable 4-door

TTI(d) test history:

Model Year	Adjusted TTI(d)	N of Tests	Side Air Bags?	Car Body Platform (Wheelbase and Other Shared Features)
1988	78.2	1		12035 FORD TAURUS 106, 86-95
1990	77.4	7		12035 FORD TAURUS 106, 86-95
1995	72.3	1		12035 FORD TAURUS 106, 86-95
1996	50.7	1		12040 FORD TAURUS 108.5, 96-
1997	57.1	1		12040 FORD TAURUS 108.5, 96-
1999	61.7	1		12040 FORD TAURUS 108.5, 96-
2000	64.7	1		12040 FORD TAURUS 108.5, 96-

Initial FMVSS 214 certification: 1996

Vehicle modifications (IR): 1996 Taurus received structural reinforcements in the body side structure, energy absorbing foam in the door panels...(no picture; not necessarily informative since this was major redesign).

Conclusions: TTI(d) was close to 72 in the first-generation Taurus/Sable, up to 1995. When these cars were redesigned and certified to FMVSS 214 in 1996, they received major structural modifications and padding, and TTI(d) decreased to about 54 in 1996-97.

Assignment: Category 1, substantial TTI(d) improvement without air bags upon original FMVSS 214 certification in 1996.

Chevrolet Corvette 2-door

TTI(d) test history:

Model Year	Adjusted TTI(d)	N of Tests	Side Air Bags?	Car Body Platform (Wheelbase and Other Shared Features)
1988	109.2	1		18051 CHEV CORVETTE Y 96.2, 84-96
1997	60.3	1		18072 CHEV CORVETTE 104.5, 97-

Initial FMVSS 214 certification: 1997

Vehicle modifications (IR): No information available.

Conclusions: The 1988 test vehicle is quite a few model years before 1996, the last year before certification. Nevertheless, specialty vehicles such as Corvette often change little from year to year between major redesigns. It is plausible that TTI(d) remained close to 109 up to 1996, and that it greatly improved, to 60, in 1997, when General Motors completely redesigned and also initially certified the Corvette. NHTSA does not have information on the vehicle modifications involved.

Assignment: Category 1, substantial TTI(d) improvement without air bags upon original FMVSS 214 certification in 1997. This make-model group is assigned to Category 1 because of the magnitude of its TTI(d) improvement, even though the vehicle modifications are unknown.

Chevrolet Cavalier/Pontiac Sunfire 2-door

TTI(d) test history:

1995	110.5	1		18066 CAVALIER/SUNFIRE J 104.1, 95-
1997	80.6	6		18066 CAVALIER/SUNFIRE J 104.1, 95-

Initial FMVSS 214 certification: 1997

Vehicle modifications (IR): 1997 Cavalier received minor structural changes (modified door beam, rear structural brace on roof rail to limit intrusion) and extensive padding.

Conclusions: TTI(d) decreased substantially from 111 in 1995-1996 to about 81 in 1997-1998, even though structural modifications appeared to have been minor, and the principal change was the addition of padding.

Assignment: Category 1, substantial TTI(d) improvement without air bags upon original FMVSS 214 certification in 1997. This is a make-model in Category 1 whose TTI(d) improvement was achieved without major structural modifications.

Chevrolet Monte Carlo 2-door

TTI(d) test history:

Model Year	Adjusted TTI(d)	N of Tests	Side Air Bags?	Car Body Platform (Wheelbase and Other Shared Features)
1995	76.6	2		18059 GM MID-SIZE W 107.5, 1988-2001
2001	62.7	1		18069 GM MID-SIZE W 110.5, 97-

Initial FMVSS 214 certification: 1995

Vehicle modifications (IR): 1995 Monte Carlo received padding and a small reinforcement at the foot of the B pillar. Note that 1995 was the first production year for the Monte Carlo. Presumably, the side structure of this new vehicle had been designed to meet FMVSS 214, and the IR merely reported final enhancements.

Conclusions: Because Monte Carlo was not produced in the years immediately before the 1995 self-certification, we cannot define "how much TTI(d) decreased in 1995." However, when Monte Carlo was remodeled in 2000, there was a moderately large decrease in TTI(d), from 77 to 63.

Assignment: Category 1, substantial TTI(d) improvement without air bags upon the 2000 redesign (and not upon the original FMVSS 214 certification in 1995).

General Motors N-Body 2-door cars[210]

TTI(d) test history:

1988	111.0	1		18054 PONT GRAND AM N 103.4, 85-98
1993	109.0	1		18054 PONT GRAND AM N 103.4, 85-98
1997	69.8	3		18054 PONT GRAND AM N 103.4, 85-98
2000	78.5	2		18068 GM MALIBU/CUTLASS 107, 97-

Initial FMVSS 214 certification: 1997

Vehicle modifications (IR): No information available.

Conclusions: TTI(d) decreased substantially from 109 in the 1993-1996 cars to 70 in 1997, upon certification. NHTSA does not have information on the vehicle modifications involved.

Assignment: Category 1, substantial TTI(d) improvement without air bags upon original FMVSS 214 certification in 1997. This make-model group is assigned to Category 1 because of the magnitude of its TTI(d) improvement, even though the vehicle modifications are unknown.

---

[210] Includes Pontiac Grand Am, Buick Skylark, and Oldsmobile Calais/Achieva/Alero.

Nissan Sentra 4-door

TTI(d) test history:

Model Year	Adjusted TTI(d)	N of Tests	Side Air Bags?	Car Body Platform (Wheelbase and Other Shared Features)
1992	91.6	1		35024 NISS SENTRA/PULSAR 95.7, 87-94
1996	66.2	1		35036 NISSAN SENTRA 99.8, 95-
1998	68.3	1		35036 NISSAN SENTRA 99.8, 95-
2002	71.3	1		35036 NISSAN SENTRA 99.8, 95-

Initial FMVSS 214 certification: 1995

Vehicle modifications (IR): 1995 Sentra received major structural changes including A pillar, B pillar and sill reinforcement, substantial padding...(not necessarily informative since this was a major redesign).

Conclusions: TTI(d) was close to 92 in the 1991-1994 Sentra. When Sentra was redesigned and certified to FMVSS 214 in 1995, it received major structural modifications and padding, and TTI(d) decreased to about 67 in 1995-1998.

Assignment: Category 1, substantial TTI(d) improvement without air bags upon original FMVSS 214 certification in 1995.

Honda Civic 2-door

TTI(d) test history:

Model Year	Adjusted TTI(d)	N of Tests	Side Air Bags?	Car Body Platform (Wheelbase and Other Shared Features)
1993	85.5	1		37023 HONDA CIVIC 103.2, 92-
1998	71.4	2		37023 HONDA CIVIC 103.2, 92-
2001	56.9	1		37023 HONDA CIVIC 103.2, 92-
2001	43.3	1	TORSO ONLY	37023 HONDA CIVIC 103.2, 92-

Initial FMVSS 214 certification: 1996

Vehicle modifications (IR): 1998 Honda Civic Coupe received substantial structural modifications including A- and B-pillar stiffeners, two cross-member reinforcements and an upgraded beam. No mention of padding. These modifications were presumably already on the 1996 and 1997 models.

Conclusions: TTI(d) decreased substantially from 86 in 1993-1995 to about 71 in 1996-1998 by major structural modifications. There may have been substantial additional improvement, even without air bags, by 2001.

Assignment: Category 1, substantial TTI(d) improvement without air bags upon original FMVSS 214 certification in 1996.

Honda Accord 2-door

TTI(d) test history:

Model Year	Adjusted TTI(d)	N of Tests	Side Air Bags?	Car Body Platform (Wheelbase and Other Shared Features)
1996	72.0	1		37026 HONDA ACCORD 106.9, 94-
1999	62.8	1		37030 HONDA ACCORD 2D COUPE 105.1, 98-
2001	46.4	1	TORSO ONLY	37030 HONDA ACCORD 2D COUPE 105.1, 98-

Initial FMVSS 214 certification: 1994

Vehicle modifications (IR): 1996 Accord received reinforcements to the A pillar, horizontal cross-members, rear wheel arch, and a redesigned beam. (The IR does not specify if these changes date back to 1994.)

Conclusions: Because we have no test results before the 1994 self-certification, we are unable to determine how much TTI(d) decreased, if at all, in 1994. However, when the Accord was remodeled in 1998, there was a moderately large improvement in TTI(d), from 72 to 63. The additional decrease to 46 in 2001 would appear to be due primarily to side air bags, not structure or padding.

Assignment: Category 1, substantial TTI(d) improvement without air bags upon the 1998 redesign (and not upon the original FMVSS 214 certification in 1994).

Honda Accord 4-door

TTI(d) test history:

Model Year	Adjusted TTI(d)	N of Tests	Side Air Bags?	Car Body Platform (Wheelbase and Other Shared Features)
1992	85.7	1		37018 HONDA ACCORD 107.1, 90-93
1994	79.1	1		37026 HONDA ACCORD 106.9, 94-
1997	74.3	1		37026 HONDA ACCORD 106.9, 94-
1998	53.6	1		37026[211] HONDA ACCORD 106.9, 94-
1999	65.1	1		37026 HONDA ACCORD 106.9, 94-
2000	44.0	1	TORSO ONLY	37026 HONDA ACCORD 106.9, 94-

Initial FMVSS 214 certification: 1994

Vehicle modifications (IR): 1994 Accord received major structural changes including A and B pillar stiffeners, horizontal cross-members, stronger beams, some padding...(not necessarily informative since this was major redesign). 1999 Accord (and presumably that includes the 1998 Accord) received a side sill extension, an additional horizontal cross-member, and additional reinforcements of the A and B pillars (again, not necessarily informative since this was a redesign).

---

[211] Honda Accord was redesigned in 1998, although it kept the same wheelbase as the 1994-1997 models.

Conclusions: Upon initial certification in 1994, TTI(d) didn't change too much – from 86 in 1992 to 77 in 1994-1997, despite the structural modifications. However, when the Accord was again remodeled in 1998, there was approximately twice as large a decrease in TTI(d), from 77 in 1994-1997 to 59 in 1998-1999. The additional decrease to 44 in 2000 would appear to be due primarily to side air bags, not structure or padding.

Assignment: Category 1, substantial TTI(d) improvement without air bags <u>upon the 1998 redesign</u> (and not upon the original FMVSS 214 certification in 1994).

Subaru Legacy 4-door

TTI(d) test history:

Model Year	Adjusted TTI(d)	N of Tests	Side Air Bags?	Car Body Platform (Wheelbase and Other Shared Features)
1995	56.2	1		48012 SUBARU LEGACY 103.5, 95-99
1997	67.1	1		48012 SUBARU LEGACY 103.5, 95-99
1998	69.2	1		48012 SUBARU LEGACY 103.5, 95-99
2000	48.2	1		48013 SUBARU LEGACY 104.3 2000-
2002	25.8	1	TORSO ONLY	48013 SUBARU LEGACY 104.3 2000-

Initial FMVSS 214 certification: 1995

Vehicle modifications (IR): 1995 Subaru Legacy wagon received substantial structural changes including stiffeners for all pillars, redesigned beams, a stronger roof, stronger seats, stronger cross-members, plus extensive padding.

Conclusions: Because we have no test results before the 1995 self-certification, we are unable to determine how much TTI(d) decreased in 1995 (even though the IR indicates considerable structural modifications). However, when Legacy was again redesigned in 2000, TTI(d) even without air bags decreased substantially from 1997-1998 levels (from 68 to 48).

Assignment: Category 1, substantial TTI(d) improvement without air bags <u>upon the 2000 redesign</u> (and not necessarily upon the original FMVSS 214 certification in 1995).

Subaru Impreza 4-door

TTI(d) test history:

1994	63.2	2		48011 SUBARU IMPREZA 99.2, 1993-2001
1995	61.1	1		48011 SUBARU IMPREZA 99.2, 1993-2001
1997	84.1	2		48011 SUBARU IMPREZA 99.2, 1993-2001
1998	72.0	1		48011 SUBARU IMPREZA 99.2, 1993-2001
2002	44.7	1		48014 SUBARU IMPREZA 99.4, 2002-

Initial FMVSS 214 certification: 1994

Vehicle modifications (IR): 1994 Impreza received padding and a strengthened seat structure and rear side door beam. Similar enhancements are described for the 1997 and 1998 Impreza 4-door.

Conclusions: Because we have no test results before the 1994 self-certification, we are unable to determine how much TTI(d) decreased, if at all, from 1993, the first year of Impreza production, to 1994. However, when Impreza was remodeled in 2002, there was a substantial improvement in TTI(d), from about 72 to 45, achieved without air bags.

Assignment: Category 1, substantial TTI(d) improvement without air bags upon the 2002 redesign (and not upon the original FMVSS 214 certification in 1994).

Toyota Corolla/Geo Prizm 4-door

TTI(d) test history:

Model Year	Adjusted TTI(d)	N of Tests	Side Air Bags?	Car Body Platform (Wheelbase and Other Shared Features)
1993	91.1	1		49030 TOYOTA COROLLA 97, 1993-2002
1997	68.1	3		49030 TOYOTA COROLLA 97, 1993-2002
1998	60.7	1		49030 TOYOTA COROLLA 97, 1993-2002
1998	48.8	1	TORSO ONLY	49030 TOYOTA COROLLA 97, 1993-2002
1999	44.9	1	TORSO ONLY	49030 TOYOTA COROLLA 97, 1993-2002
2003	53.9	1		49046 TOYOTA COROLLA 102.4, 2003-

Initial FMVSS 214 certification: 1997

Vehicle modifications (IR): 1997 Corolla received major structural changes including B pillar stiffener, horizontal cross-members, sill & roof rail strengthening, extensive padding…

Conclusions: TTI(d) decreased substantially from 91 in 1993-1996 to about 66 in 1997-1998 by major structural modifications and padding. The excellent scores for the 1998 and 1999 test vehicles appear primarily due to side air bags, not structure or padding.

Assignment: Category 1, substantial TTI(d) improvement without air bags upon original FMVSS 214 certification in 1997.

Mitsubishi Eclipse 2-door

TTI(d) test history:

Model Year	Adjusted TTI(d)	N of Tests	Side Air Bags?	Car Body Platform (Wheelbase and Other Shared Features)
1996	84.7	1		52019 MITS ECLIPSE 98.8, 95-99
1998	92.9[212]	1		52019 MITS ECLIPSE 98.8, 95-99
2000	51.6	1		52022 MITSUBISHI ECLIPSE 100.8, 2000-
2001	41.8	1	TORSO ONLY	52022 MITSUBISHI ECLIPSE 100.8, 2000-

Initial FMVSS 214 certification: End of May 1995

Vehicle modifications (IR): 1995 [sic] Mitsubishi Eclipse received reinforcements to the B pillar and cross-member, plus some padding; the IR does not specify if this applied to all MY 1995 cars or just some of them.

Conclusions: Because we have no test results before the self-certification, we are unable to determine how much TTI(d) decreased, if at all, at that time. However, when Eclipse was again redesigned in 2000, TTI(d) even without air bags decreased substantially from 1996-1999 levels (from 89 to 52).

Assignment: Category 1, substantial TTI(d) improvement without air bags <u>upon the 2000 redesign</u> (and not necessarily upon the original FMVSS 214 certification in May 1995).

Category 2: Make-model groups with no TTI(d) change or limited change

Ford Escort/Mercury Tracer 4-door

TTI(d) test history:

1983	93.5	1		12031 FORD ESCORT 94.2, 81-90
1988	98.0	1		12031 FORD ESCORT 94.2, 81-90
1995	67.4	1		41017 FORD ESCORT 98.4, 90-
1997	64.9	2		41017 FORD ESCORT 98.4, 90-

Initial FMVSS 214 certification: 1997

Vehicle modifications (IR): 1997 Escort received minor structural changes including a small reinforcement plate in the door structure, and extensive padding...

Conclusions: TTI(d) was about the same before and after certification: 67 in 1995 and 65 in 1997. This car essentially met FMVSS 214 even before it was certified, and did not receive any important structural modifications.

Assignment: Category 2, little or no TTI(d) change upon FMVSS 214 certification.

---

[212] This is the speed-adjusted TTI(d) on an NCAP test. It is not a failure on a compliance test.

Ford Probe/Mazda MX6 2-door

TTI(d) test history:

Model Year	Adjusted TTI(d)	N of Tests	Side Air Bags?	Car Body Platform (Wheelbase and Other Shared Features)
1993	82.5	1		41021 MAZDA 626/PROBE 102.9, 93-97
1997	80.2	1		41021 MAZDA 626/PROBE 102.9, 93-97

Initial FMVSS 214 certification: 1997

Vehicle modifications (IR): 1997 Ford Probe added padding to the door panels and had the window regulator motor revised.

Conclusions: TTI(d) in 1997 (80) was about the same as in 1993-1996 (83). The test results are consistent with the relatively minor modifications described in the IR.

Assignment: Category 2, little or no TTI(d) change upon FMVSS 214 certification.

Lincoln Town Car 4-door

TTI(d) test history:

1988	40.0	1		12030 LINCOLN TOWN CAR 117.7, 80-
1994	62.5	1		12030 LINCOLN TOWN CAR 117.7, 80-
1999	46.3	1	COMBINATION	12030 LINCOLN TOWN CAR 117.7, 80-

Initial FMVSS 214 certification: 1994

Vehicle modifications (IR): 1994 Lincoln Town Car had minor changes related to FMVSS 214 (3 electrical connectors moved from the inner to the outer door panel).

Conclusions: TTI(d) did not decrease upon certification in 1994. We have no explanation for the observed increase from 40 in 1988 to 63 in 1994, considerably more than the typical test-to-test variation, but in any case it is not a decrease.

Assignment: Category 2, little or no TTI(d) change upon FMVSS 214 certification.

Mercury Grand Marquis/Ford Crown Victoria 4-door

TTI(d) test history:

1992	41.2	1		12028 CROWN VIC/GRAND MARQUIS, 79-
1997	51.0	2		12028 CROWN VIC/GRAND MARQUIS, 79-

Initial FMVSS 214 certification: 1994

Vehicle modifications (IR): 1997 Ford Crown Victoria did not receive any modifications to meet 214 and is a carryover from MY 1995 [sic].

Conclusions: These cars easily met FMVSS 214 even before 1994. We have no evidence that they were modified or that TTI(d) decreased upon self-certification in 1994.

Assignment: Category 2, little or no TTI(d) change upon FMVSS 214 certification.

Buick LeSabre/Oldsmobile 88/Pontiac Bonneville 4-door

TTI(d) test history:

Model Year	Adjusted TTI(d)	N of Tests	Side Air Bags?	Car Body Platform (Wheelbase and Other Shared Features)
1988	77.6	2		18052 GM LUX C/FULL H 110.8, 84-99
1995	72.2	1		18052 GM LUX C/FULL H 110.8, 84-99
1997	69.1	4		18052 GM LUX C/FULL H 110.8, 84-99
1998	62.3	2		18052 GM LUX C/FULL H 110.8, 84-99
2000	54.1	2	TORSO ONLY	18067 LeSABRE/SEVILLE 112.2, 98-

Initial FMVSS 214 certification: 1997

Vehicle modifications (IR): 1997 LeSabre received thorax and pelvic padding, armrest force deflection characteristics modified, and the side ring structure redesigned

Conclusions: TTI(d) was about the same before and after certification: 72 in 1995-1996 and 67 in 1997-1998. This car essentially met FMVSS 214 even before it was certified, and received padding, but no important structural modifications.

Assignment: Category 2, little or no TTI(d) change upon FMVSS 214 certification.

Chevrolet Caprice/Buick Roadmaster 4-door sedan

TTI(d) test history:

| 1988 | 50.3 | 2 | | 18039 GM FULLSIZED SEDAN 116, 77-96 |
| 1994 | 56.3 | 3 | | 18039 GM FULLSIZED SEDAN 116, 77-96 |

Initial FMVSS 214 certification: 1994

Vehicle modifications (IR): 1994 Buick Roadmaster did not receive any modifications to meet FMVSS 214.

Conclusions: These cars easily met FMVSS 214 even before 1994. They were not modified upon self-certification in 1994.

Assignment: Category 2, little or no TTI(d) change upon FMVSS 214 certification.

Chevrolet Camaro/Pontiac Firebird 2-door

TTI(d) test history:

Model Year	Adjusted TTI(d)	N of Tests	Side Air Bags?	Car Body Platform (Wheelbase and Other Shared Features)
1995	77.3	3		18049 CHEV CAMARO F 101, 1982-2002
1997	66.8	5		18049 CHEV CAMARO F 101, 1982-2002

Initial FMVSS 214 certification: 1995

Vehicle modifications (IR): 1995 Camaro unchanged from 1994 except padding added to rear armrests.

Conclusions: These cars were not modified to improve front-seat TTI(d) upon self-certification in 1995.

Assignment: Category 2, little or no TTI(d) change upon FMVSS 214 certification.

Chevrolet Cavalier/Pontiac Sunfire 4-door

TTI(d) test history:

1986	81.2	1		18048 GM COMPACT J CARS 101.2, 82-94
1987	82.2	1		18048 GM COMPACT J CARS 101.2, 82-94
1997	74.3	2		18066 CAVALIER/SUNFIRE J 104.1, 95-
1998	82.0	2		18066 CAVALIER/SUNFIRE J 104.1, 95-

Initial FMVSS 214 certification: 1997

Vehicle modifications (IR): 1998 Chevrolet Cavalier 4-door sedan received extensive padding, no structural modifications. Presumably that's retroactive to 1997.

Conclusions: TTI(d) probably was close to 80 throughout 1986-1998. While it is not impossible that TTI(d) became much worse from 1986-1987 to 1993-1996 and then greatly improved in 1997, it is hardly likely.

Assignment: Category 2, little or no TTI(d) change upon FMVSS 214 certification.

Chevrolet Lumina 4-door

TTI(d) test history:

Model Year	Adjusted TTI(d)	N of Tests	Side Air Bags?	Car Body Platform (Wheelbase and Other Shared Features)
1992	65.6	1		18059 GM MID-SIZE W 107.5, 1988-2001
1995	61.7	1		18059 GM MID-SIZE W 107.5, 1988-2001
1997	55.6	2		18059 GM MID-SIZE W 107.5, 1988-2001

Initial FMVSS 214 certification: 1995

Vehicle modifications (IR): 1995 Lumina received negligible structural change consisting of a redesigned B-pillar to rocker panel joint, no front-seat padding. 1997 Lumina is apparently identical to the 1995.

Conclusions: TTI(d) was about the same before and after certification: 66 in 1992-1994 and 62 in 1995 (58 if you average the 1995-1997 tests). This car essentially met FMVSS 214 even before it was certified, and did not receive any important structural modifications.

Assignment: Category 2, little or no TTI(d) change upon FMVSS 214 certification.

General Motors N-Body 4-door cars[213]

TTI(d) test history:

1988	73.5	1		18054 PONT GRAND AM N 103.4, 85-98
1996	64.0	1		18054 PONT GRAND AM N 103.4, 85-98
1997	71.8	5		18054 PONT GRAND AM N 103.4, 85-98
1997	79.2	2		18068 GM MALIBU/CUTLASS 107, 97-
1999	71.9	2		18068 GM MALIBU/CUTLASS 107, 97-
2000	70.4	1		18068 GM MALIBU/CUTLASS 107, 97-
2002	62.3	1		18068 GM MALIBU/CUTLASS 107, 97-

Initial FMVSS 214 certification: 1997

Vehicle modifications (IR): 1997 Grand Am received thorax and pelvic padding, but no structural changes.

Conclusions: TTI(d) was close to 70 throughout 1988-2000, and did not decrease upon certification in 1997. The cars did not receive structural modifications at that time.

Assignment: Category 2, little or no TTI(d) change upon FMVSS 214 certification.

---

[213] Includes Pontiac Grand Am, Buick Skylark, Chevrolet Malibu (1997-) and Oldsmobile Calais/Achieva/Alero.

Saturn 4-door

TTI(d) test history:

Model Year	Adjusted TTI(d)	N of Tests	Side Air Bags?	Car Body Platform (Wheelbase and Other Shared Features)
1995	68.8	1		18062 SATURN 102.4, 1991-2002
1996	67.0	3		18062 SATURN 102.4, 1991-2002
1997	71.1	2		18062 SATURN 102.4, 1991-2002
2000	72.9	3		18062 SATURN 102.4, 1991-2002

Initial FMVSS 214 certification: 1996

Vehicle modifications (IR): 1996 Saturn received major structural reinforcements in the A, B and C pillars, cross-members, beams, rocker panels, plus extensive padding. 1997 Saturn is unchanged from 1996.

Conclusions: Saturn had nearly the same TTI(d) before certification (69 in 1995) and afterwards (69 in 1996-98), notwithstanding the major structural reinforcements described in the IR.

Assignment: Category 2, little or no TTI(d) change upon FMVSS 214 certification.

Honda Civic 4-door

TTI(d) test history:

Model Year	Adjusted TTI(d)	N of Tests	Side Air Bags?	Car Body Platform (Wheelbase and Other Shared Features)
1995	53.5	1		37023 HONDA CIVIC 103.2, 92-
1997	60.7	2		37023 HONDA CIVIC 103.2, 92-
2001	47.1	1		37023 HONDA CIVIC 103.2, 92-
2001	46.7	1	TORSO ONLY	37023 HONDA CIVIC 103.2, 92-

Initial FMVSS 214 certification: 1996

Vehicle modifications (IR): 1997 Civic received major structural reinforcements in the A and B pillars, cross-members, beams, sills, plus some padding. IR does not say these are carryovers from 1996, but presumably they were.

Conclusions: TTI(d) did not substantially decrease in 1996; in fact, test result for 1997 (61) is slightly higher than in 1995 (54). TTI(d) without air bags in 2001 (47) is somewhat lower than in 1997, but not that much lower than 1995.

Assignment: Category 2, little or no TTI(d) change upon FMVSS 214 certification.

Category 3: Make-model groups that may have received major new structure upon FMVSS 214 certification, TTI(d) change unknown

Mazda Protégé 4-door

TTI(d) test history:

Model Year	Adjusted TTI(d)	N of Tests	Side Air Bags?	Car Body Platform (Wheelbase and Other Shared Features)
1997	62.3	1		18065 GM AURORA/RIVIERA G 113.8, 95-
2001	46.0	1	TORSO ONLY	18065 GM AURORA/RIVIERA G 113.8, 95-

Initial FMVSS 214 certification: 1995

Vehicle modifications (IR): 1995 Mazda Protégé received major structural changes including B pillar stiffener, sill reinforcement, stronger cross-members, plus extensive padding.

Conclusions: Because we have no test results before the 1995 self-certification, we are unable to determine how much TTI(d) decreased, even though the IR indicates major structure modifications. The fine score for the 2001 test vehicle appears primarily due to side air bags.

Assignment: Category 3, may have received major new structure upon FMVSS 214 certification in 1995, TTI(d) change unknown.

Toyota Celica 2-door

TTI(d) test history:

1997	59.4	1		49033 TOYOTA CELICA 99.9, 94-99
2001	67.9	1		49039 TOYOTA CELICA 102.3, 2000-

Initial FMVSS 214 certification: 1996

Vehicle modifications (IR): 1996 Celica received a reinforced B pillar and center cross member, plus extensive padding.

Conclusions: Because we have no test results before the 1996 self-certification, we are unable to determine how much TTI(d) decreased in 1996, even though the IR indicates apparently major structure modifications.

Assignment: Category 3, may have received major new structure upon FMVSS 214 certification in 1996, TTI(d) change unknown.

Toyota Tercel 4-door

TTI(d) test history:

Model Year	Adjusted TTI(d)	N of Tests	Side Air Bags?	Car Body Platform (Wheelbase and Other Shared Features)
1995	61.4	1		49025 TOYOTA TERCEL 93.7, 87-98
1997	64.0	1		49025 TOYOTA TERCEL 93.7, 87-98

Initial FMVSS 214 certification: 1995

Vehicle modifications (IR): 1995 Toyota Tercel received major structural changes including B-pillar stiffener, a second beam, stronger cross-members, plus extensive padding.

Conclusions: Because we have no test results before the 1995 self-certification, we are unable to determine how much TTI(d) decreased, even though the IR indicates major structure modifications.

Assignment: Category 3, may have received major new structure upon FMVSS 214 certification in 1995, TTI(d) change unknown.

Toyota Camry 4-door

TTI(d) test history:

1994	64.9	1		49028 TOYOTA CAMRY 103.1, 92-96
1997	63.0	1		49036 TOYOTA CAMRY 105.1, 97-
1998	49.5	1	TORSO ONLY	49036 TOYOTA CAMRY 105.1, 97-
1999	49.0	1	TORSO ONLY	49036 TOYOTA CAMRY 105.1, 97-
2002	64.0	3		49035 TOYOTA AVALON/CAMRY 107.1, 95-

Initial FMVSS 214 certification: 1994

Vehicle modifications (IR): 1994 Camry received major structural changes including A and B pillar stiffeners, horizontal cross-members, extensive padding… 1998 Camry received reinforced rocker panel, B-pillar, front floor cross member, roof side rail, horizontal roof reinforcement, an additional side door beam, and padding in two places. (It is not clear whether all of these are over and above the 1994 changes, or merely a recapitulation of those changes; at a minimum, the horizontal roof reinforcement and additional side door beam are new. Presumably, these changes were already present in the redesigned 1997 Camry.)

Conclusions: Because we have no test results before the 1994 self-certification, we are unable to determine how much TTI(d) decreased, even though the IR suggests major structure modifications. The 1997 remodeling did not substantially improve TTI(d) from the 1994-1996 level (63 vs. 65), and the 2002 remodeling didn't either. The excellent scores for the 1998 and 1999 test vehicles appear primarily due to side air bags, not structure or padding.

Assignment: Category 3, may have received major new structure upon FMVSS 214 certification in 1994, TTI(d) change unknown.

Mitsubishi Galant 4-door

TTI(d) test history:

Model Year	Adjusted TTI(d)	N of Tests	Side Air Bags?	Car Body Platform (Wheelbase and Other Shared Features)
1994	63.6	1		52018 MITS GALANT 103.7, 94-
1995	75.0	1		52018 MITS GALANT 103.7, 94-
1997	64.6	1		52018 MITS GALANT 103.7, 94-
1999	46.1	2		52018 MITS GALANT 103.7, 94-
2001	65.8	1		52018 MITS GALANT 103.7, 94-

Initial FMVSS 214 certification: 1994

Vehicle modifications (IR): 1994 Galant received reinforcements to the B pillar, horizontal cross-members, sills, and a redesigned beam, plus extensive padding.

Conclusions: Because we have no test results before the 1994 self-certification, we are unable to determine how much TTI(d) decreased in 1994, even though the IR indicates major structure modifications. The TTI(d) history from 1994 onwards shows larger than usual test-to-test variations, ranging as high as 75 and as low as 46, but returning to 66 in 2001. From this, we cannot conclude that TTI(d) was substantially and permanently improved at any time after 1994, either.

Assignment: Category 3, may have received major new structure upon FMVSS 214 certification in 1994, TTI(d) change unknown.

Category 4: Other make-model groups with unknown TTI(d) change upon FMVSS 214 certification

Ford Thunderbird/Mercury Cougar 2-door

TTI(d) test history:

1995	70.5	1		12037 FORD T-BIRD 113, 89-98
1997	69.5	1		12037 FORD T-BIRD 113, 89-98
2002	37.6	1	COMBINATION	12044 FORD T-BIRD 107.2, 2002-

Initial FMVSS 214 certification: 1995

Vehicle modifications (IR): 1995 Thunderbird received structural reinforcement in the body side structure, padding in the door trim panel (no picture, unclear if these were moderate or minor changes).

Conclusions: Because we have no test results before the 1995 self-certification, we are unable to determine how much TTI(d) decreased, if at all, in 1995. The outstanding test performance of the 2002 Thunderbird is to a large extent due to side air bags with head protection.

Assignment: Category 4, unknown TTI(d) change upon FMVSS 214 certification.

Buick Park Avenue/Oldsmobile 98 4-door

TTI(d) test history:

Model Year	Adjusted TTI(d)	N of Tests	Side Air Bags?	Car Body Platform (Wheelbase and Other Shared Features)
1997	62.3	1		18065 GM AURORA/RIVIERA G 113.8, 95-
2001	46.0	1	TORSO ONLY	18065 GM AURORA/RIVIERA G 113.8, 95-

Initial FMVSS 214 certification: 1997

Vehicle modifications (IR): No information available.

Conclusions: Because we have no test results before the 1997 certification, we are unable to determine how much TTI(d) decreased, if at all, in 1997. The fine test performance of the 2001 Park Avenue is to a large extent due to side air bags.

Assignment: Category 4, unknown TTI(d) change upon FMVSS 214 certification.

Cadillac DeVille 4-door

TTI(d) test history:

Model Year	Adjusted TTI(d)	N of Tests	Side Air Bags?	Car Body Platform (Wheelbase and Other Shared Features)
1994	53.7	2		18064 CADI DeVILLE K 113.8, 94-99
1997	45.4	2	TORSO ONLY	18064 CADI DeVILLE K 113.8, 94-99
2000	50.0	1	TORSO ONLY	18074 CADILLAC DeVILLE 115.3 2000-

Initial FMVSS 214 certification: 1994

Vehicle modifications (IR): The manufacturer states that 1994 Cadillac DeVille "did not receive any modifications" to meet 214. However, it is not clear what that means, because DeVille received a major redesign in 1994 and modifications could have been "built into" rather than "added onto" the new design.

Conclusions: Because we have no test results before the 1994 self-certification, we are unable to determine how much TTI(d) decreased, if at all. The excellent scores for the 1997 and 2000 test vehicles appear primarily due to side air bags, not structure or padding.

Assignment: Category 4, unknown TTI(d) change upon FMVSS 214 certification.

Geo Metro/Suzuki Swift 2-door

TTI(d) test history:

Model Year	Adjusted TTI(d)	N of Tests	Side Air Bags?	Car Body Platform (Wheelbase and Other Shared Features)
1995	77.0	1		53004 GEO METRO 93.1, 1989-2001
1996	82.0	1		53004 GEO METRO 93.1, 1989-2001

Initial FMVSS 214 certification: 1995

Vehicle modifications (IR): 1996 Geo Metro 3-door is a carryover from MY 1995; no changes for 1996 [sic – from this it is unclear if anything was done in 1995, the actual year of initial certification].

Conclusions: Because we have no test results before the 1995 self-certification, we are unable to determine how much TTI(d) decreased, if at all, in 1995. The IR does not indicate what sort of modifications, if any, were implemented in 1995.

Assignment: Category 4, unknown TTI(d) change upon FMVSS 214 certification.

Saturn 2-door

TTI(d) test history:

1997	77.7	2		18062 SATURN 102.4, 1991-2002
1998	67.7	1		18062 SATURN 102.4, 1991-2002

Initial FMVSS 214 certification: 1997

Vehicle modifications (IR): 1997 Saturn SC2 received extensive reinforcements and a high-strength beam on the rear quarter panel; in the door area they added padding but it's not clear there were any structural modifications.

Conclusions: Because we have no test results before the 1997 certification, we are unable to determine how much TTI(d) decreased, if at all, in 1997. It is also unclear if Saturn coupe received minor or moderate structural changes in 1997.

Assignment: Category 4, unknown TTI(d) change upon FMVSS 214 certification.

Nissan Maxima 4-door

TTI(d) test history:

Model Year	Adjusted TTI(d)	N of Tests	Side Air Bags?	Car Body Platform (Wheelbase and Other Shared Features)
1995	63.6	1		35034 NISSAN MAXIMA 106.3, 95-99
1997	53.1	1		35034 NISSAN MAXIMA 106.3, 95-99
1998	53.1	1	TORSO ONLY	35034 NISSAN MAXIMA 106.3, 95-99
2000	60.7	1		35039 NISSAN MAXIMA 108.3, 2000-
2000	44.4	1	COMBINATION	35039 NISSAN MAXIMA 108.3, 2000-
2001	53.5	1		35039 NISSAN MAXIMA 108.3, 2000-

Initial FMVSS 214 certification: 1995

Vehicle modifications (IR): 1998 Maxima received reinforcements of the sill and the A- and B-pillars plus padding.

Conclusions: Because we have no test results before the 1995 self-certification, we are unable to determine how much TTI(d) decreased, if at all, in 1995. When Maxima was again redesigned in 2000, TTI(d) (without air bags) was about the same as in the 1995-1999 version.

Assignment: Category 4, unknown TTI(d) change upon FMVSS 214 certification.

Nissan Altima 4-door

TTI(d) test history:

1997	70.2	1		35032 NISSAN ALTIMA 103.1, 1993-2001
1998	67.7	2		35032 NISSAN ALTIMA 103.1, 1993-2001
2002	65.1	1		35040 NISSAN ALTIMA 110.2, 2002-

Initial FMVSS 214 certification: 1997

Vehicle modifications (IR): 1997 Altima had certain door structure, body structure in the area of the door and interior components redesigned (no picture, unclear if these were moderate or minor changes).

Conclusions: Because we have no test results before the 1997 certification, we are unable to determine how much TTI(d) decreased, if at all. It is also unclear if Altima received minor or moderate structural changes in 1997.

Assignment: Category 4, unknown TTI(d) change upon FMVSS 214 certification.

Mazda 626 4-door

TTI(d) test history:

Model Year	Adjusted TTI(d)	N of Tests	Side Air Bags?	Car Body Platform (Wheelbase and Other Shared Features)
1983	105.1	1		41010 MAZDA 626 98.8 FWD, 83-87
1996	60.3	1		41021 MAZDA 626/PROBE 102.9, 93-97
1997	71.2	1		41021 MAZDA 626/PROBE 102.9, 93-97
1998	59.8	2		41025 MAZDA 626 105.1, 98-

Initial FMVSS 214 certification: 1996

Vehicle modifications (IR): 1996 Mazda 626 received minor-moderate structural changes including small reinforcements in the B pillar and some door panels, plus extensive padding.

Conclusions: We are unable to determine how much TTI(d) decreased, if at all, in 1996 (the 1983 test result is too old and we have no information on the next two generations of pre-standard 626's). It is unclear if the 1998 redesign substantially improved TTI(d) from the 1996-1997 level or left it about the same.

Assignment: Category 4, unknown TTI(d) change upon FMVSS 214 certification.

www.ingramcontent.com/pod-product-compliance
Lightning Source LLC
Chambersburg PA
CBHW081446170526
45166CB00008B/2337